# River Flow and Reservoir Management

# River Flow and Reservoir Management

*Edited by:*

**Sujata Nagnath Mustapure**

Delve
PUBLISHING

www.delvepublishing.com

**River Flow and Reservoir Management**

*Sujata Nagnath Mustapure*

**Delve Publishing**
224 Shoreacres Road
Burlington, ON L7L 2H2
Canada
www.delvepublishing.com
Email: orders@arclereducation.com



ISBN: 978-1-77469-020-8 (Hardcover)

# ABOUT THE EDITOR

**Dr. Sujata Nagnath Mustapure** is a teaching associate, Department of Electrical & Other Energy sources, College of agricultural engineering & Technology, Vasantrao Naik Marathwada Krishi Vidhyapeet, Parbhani. She obtained her B.Tech (agril. Engg.) in 2010 from Vasantrao Naik Marathwada Krishi Vidhyapeet, Parbhani. Gold medal in M.Tech (Agril. Engg.), 2014 from Dr. Panjabrao Deshmukh Kríshi Vidyapeeth, Akola and Ph.D in Renewable energy engineering in 2019 from Maharana Pratap University of Agriculture and Technology, Udaipur. She has also worked as assistant professor in College of Agricultural, Naigoan, Vasantrao Naik Marathwada Krishi Vidhyapeet, Parbhani. She has authored several national and international publications.

# TABLE OF CONTENTS

*List of Figures* ........ *xi*

*Preface* ........ *xv*

**Introduction** ........ **1**

**Chapter 1 Advanced Reservoir Management** ........ **3**

1.1 Introduction ........ 4

1.2 Flow Dynamics ........ 5

1.3 Permeability ........ 7

1.4 Flow Regimes ........ 8

**Chapter 2 River Flow** ........ **13**

2.1 Introduction ........ 14

2.2 Hydrological Conditions ........ 16

2.3 Fluvial Geomorphology ........ 17

2.4 Rills ........ 19

2.5 Streams ........ 22

2.6 Climate Associations ........ 32

**Chapter 3 Environmental Management of River Basin** ........ **35**

3.1 Introduction ........ 36

3.2 Flooding ........ 37

3.3 GCMs ........ 45

**Chapter 4 Introduction to Reservoir Engineering** ........ **49**

4.1 Introduction ........ 50

4.2 Reservoir Engineering ........ 50

4.3 Design ........ 53

4.4 Mineral Deposits ........ 54

4.5 Reservoir Rocks ..... 56
4.6 Logging Tools ..... 56
4.7 Permeability ..... 57
4.8 Well Logging ..... 60

**Chapter 5 Introduction to Water Resources Engineering ..... 67**
5.1 Introduction ..... 68
5.2 Closed Conduits ..... 71
5.3 Evapotranspiration ..... 76
5.4 Rivers ..... 80
5.5 Stream Orders ..... 81
5.6 Forecasting ..... 82
5.7 Artificial Intelligence ..... 85
5.8 Models ..... 86

**Chapter 6 River Hydrology ..... 91**
6.1 Introduction ..... 92
6.2 Hydraulics ..... 94
6.3 Floodplains ..... 97
6.4 Braided Rivers ..... 98
6.5 Watershed ..... 105
6.6 Sediment Transportation ..... 110

**Chapter 7 River Pollution ..... 115**
7.1 Introduction ..... 116
7.2 Physical Remediation ..... 119
7.3 Chemical Remediation ..... 120
7.4 Dissolved Oxygen ..... 124
7.5 Agriculture ..... 128

**Chapter 8 Reservoir Pollution ..... 131**
8.1 Introduction ..... 132
8.2 Chemical Characteristics ..... 135
8.3 Microbiology ..... 140
8.4 Water Quality ..... 141
8.5 Operation ..... 142

**Chapter 9 Geomorphology of Rivers** ........ **149**
9.1 Introduction ........ 150
9.2 Rivers ........ 150
9.3 Environmental Legislation ........ 153
9.4 Bedrock Channels ........ 156

**Chapter 10 River Ecosystems** ........ **177**
10.1 Introduction ........ 178
10.2 Corridors ........ 178
10.3 River Corridors ........ 180
10.4 Aquatic Ecosystems ........ 181
10.5 Riparian Ecosystems ........ 182
10.6 Ecological Functions ........ 184
10.7 Habitat Fragmentation ........ 188

**Chapter 11 Mitigation of Floods** ........ **191**
11.1 Introduction ........ 192
11.2 Climate Change Perspectives ........ 195
11.3 Flood Characteristics ........ 196
11.4 Hydroinformatics ........ 196
11.5 Knowledge Management ........ 198
11.6 Geopolitics ........ 205
11.7 Environmental Impact Assessment ........ 211

**Index** ........ **213**

# LIST OF FIGURES

**Figure 1.1** Flow dynamics
**Figure 1.2** Darcy's law
**Figure 1.3** Permeability
**Figure 1.4** Flow regimes
**Figure 2.1** River Flow
**Figure 2.2** Climate change patterns
**Figure 2.3** Hydrological Cycle
**Figure 2.4** Fluvial Geomorphology
**Figure 2.5** Rills
**Figure 2.6** Streams
**Figure 2.7** Antidunes
**Figure 2.8** Ripples
**Figure 2.9** Dunes
**Figure 2.10** Groundwater flow
**Figure 2.11** Soil erosion
**Figure 2.12** Gully erosion
**Figure 2.13** Sediment transport
**Figure 2.14** Channel geometry
**Figure 2.15** Meandering of rivers
**Figure 3.1** Flooding
**Figure 3.2** Floodplains
**Figure 3.3** Integrated watershed management
**Figure 3.4** Management of watersheds
**Figure 3.5** Sub-watersheds
**Figure 3.6** Micro watersheds
**Figure 3.7** Stream order
***Figure 3.8*** Freeze and Harlan proposed hydrological models
**Figure 3.9** GCM

**Figure 3.10** Weather research and forecasting
**Figure 4.1** Reservoir engineering
**Figure 4.2** Reservoir boundary
**Figure 4.3** Mineral deposits
**Figure 4.4** Rock permeability
**Figure 4.5** Well logging
**Figure 4.6** Resistivity logs
**Figure 4.7** Acousticlogs
**Figure 4.8** Gamma ray logs
**Figure 4.9** Caliperlogs
**Figure 5.1** Water Resource Engineering
**Figure 5.2** Infiltration of water
**Figure 5.3** Impoundments
**Figure 5.4** Closed conduits
**Figure 5.5** Condensation of clouds
**Figure 5.6** Bergeron-Findeisen process
**Figure 5.7** Hortonian runoff
**Figure 5.8** Potential evapotranspiration
**Figure 5.9** Artificial Intelligence
**Figure 6.1** Braided rivers
**Figure 6.2** Bioaccumulation
**Figure 7.1** River pollution
**Figure 7.2** Riverine ecosystems
**Figure 7.3** Algal blooms
**Figure 7.4** Dissolved oxygen
**Figure 8.1** Phytoplankton
**Figure 8.2** Oscillatoria
**Figure 8.3** Cyanobacteria
**Figure 8.4** Zooplankton
**Figure 9.1** India-Asia Plates
**Figure 9.2** Mississippi
**Figure 9.3** Amazon
**Figure 9.4** GIS tools

**Figure 9.5** Digital Elevation Models
**Figure 9.6** Bedrock channels
**Figure 9.7** Fluvial Hydraulics
**Figure 10.1** Stream corridors
**Figure 10.2** River corridors
**Figure 10.3** Ecological functions
**Figure 10.4** Habitat Fragmentation
**Figure 11.1** Geostatistical tools
**Figure 11.2** Flood disasters
**Figure 11.3** Encroachment of floodplains
**Figure 11.4** Hydroinformatics
**Figure 11.5** Geopolitics
**Figure 11.6** Ohio river
**Figure 11.7** Missouri river

# PREFACE

The rivers have brought us diversity in the foods and helped us to develop several civilizations that made us now. The rivers are considered as divine with several opportunities for humans to explore. The sensibilities and delicacies that we enjoy have derived from these prestigious water bodies. Most of the river bodies are subjected to several external and internal forces that alter their existence with geological time. Most of the civilizations related to the riverine ecosystems and this was proved by several archaeologists and historians. The water is fed into the rivers through several sources and this will bring culmination of minerals and other essential elements in them. The rivers will be obstructed with the aid of reservoirs to store water that otherwise reaches the oceans and may damage the local landscape and biota living on it. The reservoirs can support diverse biotic species that demand several interactions among the surface and submerged components.

The eutrophication of the rivers can lead to the pollution of the reservoirs and this will effectively reduce the water quality that is already stored. The internal structures of the reservoirs can affect the biotic components and several microbes can act upon aquatic life. The sediments get dissolved in the stored water and in some cases based on the flow rates, we can expect high sedimentation rates. The precipitation patterns and surface runoff can adversely affect the quality of the reservoirs. Advanced reservoir management is necessary to cope up with the changes that might occur in these massive water bodies and mitigate the uneven reactions that might occur. The pollution of rivers, streams and reservoirs is explained in this book. The source and sink model of the river ecosystems are illustrated accordingly. The design parameters of the reservoir along with its operation and maintenance are described. The role of riparian systems in loading the rivers with sediments and pollutants was discussed in detail.

The terrestrial ecosystems that are dominated by vegetation can reduce the surface runoff and hence water that reaches the reservoirs can be limited. This can also affect the local ecosystems downstream and upstream of reservoirs. This book emphasizes the role of the water engineers and managers who regulate and oversee the operation of the reservoirs. The role of flood management authorities in managing the public in case of disastrous events is explained. The river ecosystems and their components within are explained with a special focus on external and internal environmental stresses.

This book presumes that the readers have certain basic knowledge of river ecosystems and reservoirs. This part of literature can provide an impetus towards the management of rivers and reservoirs in diverse landscapes. This book can be used as a guide for hydrologists and students who want to pursue water-related courses at various levels.

# Introduction

Most of the river ecosystems are subjected to restoration and this can be due to increased pollution loads existing in them. The siltation and sedimentation of the rivers have become a common cause for most of the water bodies that are natural and artificially built. The restoration process can be based on the integrated use of biological processes along with chemical and physical factors. Anthropogenic activities have created several problems with the riverine habitats and reservoir biological communities. The freshwater ecosystems can play an important role in sustaining most of the biological life and it is being polluted by the industries causing a cascade of pollution loading events. The downstream regions of the rivers are being affected badly and so the marine environment.

The reservoir engineering has brought us a certain hope to withstand the forces caused by the rivers during heavy precipitation. The devastating flood situations are being averted due to dams and associated reservoirs thereof. Some of the hydrodynamics were discussed in this book. The role of information technology in early warning systems was discussed in this part of literature. The flood mitigation and monitoring techniques along with the role of the community-based NGOs and public authorities was emphasized. Geomorphology of the rivers was explained with a special focus on the shaping of the rivers due to external and internal stresses.

Reservoir engineering and water resource engineering have provided us an excellent knowledge base to know the structures built across the river basins. Latest techniques like remote sensing and GIS can be used in the mapping of the water related projects. We can use several mapping techniques in investigating the actual conditions of the local water bodies. It is crucial to know the pollutant load at all the segments of the river. This book can illustrate the techniques needed in the planning and design of the reservoirs.

Flood mitigation techniques were emphasized in this book so that the students can easily understand the concept in a simple manner. The pollution of the lakes, streams, rivers, and reservoirs was discussed. The role of the engineers in managing the water bodies and pollution control was also discussed.

Chapter 1

# Advanced Reservoir Management

## CONTENTS

1.1 Introduction .... 4
1.2 Flow Dynamics .... 5
1.3 Permeability .... 7
1.4 Flow Regimes .... 8

# 1.1 INTRODUCTION

The flow that is through porous media can be classified as a complex process and we can't simply explain the whole process with simple conduits and pipes. It is easier to calculate the diameter and length of a specific pipe. We can also measure the flow capacity if we know the pressure. It is not the case in porous media as we cannot demarcate flow paths. This can be a hard task to endure. There were several attempts made to understand this concept. There are two primordial fronts through which this research progresses. They are experimental and analytical fronts. Most of the research communities are using the experimental front to understand the flow patterns and fluid behavior. These studies can be common in sand packs. On the analysis of these parameters and values, we can formulate laws according to the demands. The main characteristics of a reservoir may include the number of fluids flowing in the reservoir, the geometry of the reservoir, flow regimes and fluid types (Holm., 1998).

The compressibility coefficient can be evaluated in isothermal conditions. This can regulate the fluid type of the reservoir. We can classify these fluids into three main groups. They are compressible fluids, slightly compressible and incompressible fluids. If a fluid's density or volume does not vary with the pressure changes, then we can call such a type of fluid incompressible fluids. The slightly compressed fluids present minor changes in density or volume according to the pressure variations. The water systems and crude oils can be classified under this category. Flow regimes are classified based on the reservoir pressure and flow behavior. They are a steady-state flow regime, unsteady state flow, and pseudo steady-state flow. In the case of steady-state flow, identified pressure at any point in the reservoir can be constant and it may never change with time. The steady-state flow conditions occur only if the Reservoir is completely recharged. The pressure maintenance operations and strong aquifer can play a major role in maintaining steady-state flow conditions of the reservoir. The unsteady state flow regimes can be called transient flow. This can be described as a fluid flowing condition. In this case, the change of pressure with time will not be zero. Pseudo steady-state flow occurs when there is a continuous decline in the pressure at various locations of the reservoir along with the time. The pseudo steady-state flow can be called as semi steady-state flow. This can be seen in slightly compressible fluids (Lanl., 2000).

## 1.2 FLOW DYNAMICS

The reservoirs may contain irregular boundaries. It is essential to use numerical simulators in this case. The flow geometry can be represented by radial flow, linear flow, spherical and hemispherical flow (Figure 1.1). if there are no noticeable heterogeneities in the reservoir then the flow can be radial. The radial flow can be observed far away from the wellbore. The radial flow concept can be used to describe the flow characteristics at the wellbore. The fluids can move towards the well from different directions. Linear flow occurs if there are parallel flow paths. In this case, the fluids will flow in a single direction. For linear flow, there will be constant cross-sectional area. This spherical and hemispherical flow can be based on the configuration of the wellbore. The mathematical expressions may be used in the introduction of Reservoir behavior and its volumetric performance. The mathematical expressions must be designed carefully as there might be several mobile fluids flowing into the reservoir. We can observe single-phase flow in the case of gas water and oil Wells. We can also observe two-phase flow and three-phase flow. If the number of fluids increases then obtaining pressure data will be a big problem. We can use fluid flow equations to describe the behavior of fluids in the reservoir and it will be based on several variables. We can combine the transport equation and conservation of mass equation in the fluid flow equation models. Mostly the flow equations can be based on Darcy's law (Ronaldo et al., 2006).

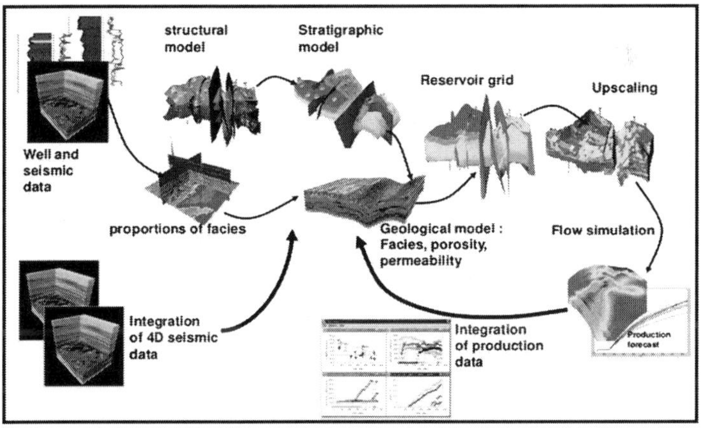


**Figure 1.1** Flow dynamics.

**Source:** *http://www.oil-gasportal.com/reservoir-management/integrated-reservoir-modeling/*

Darcy's law (Figure 1.2) can be considered as the fundamental law that can explain fluid motion, especially in the Porous media. According to this law the velocity of the homogeneous fluid may be proportional to the pressure gradient. This is true in the porous medium. The velocity can be inversely proportional to the viscosity of the fluids. In this equation, we can consider the total cross-sectional area of the rock, volumetric flow rate and apparent velocity. Darcy's law can be applied in certain conditions only. They are homogeneous formation, incompressible fluids, steady-state flow and laminar flow or viscous flow.

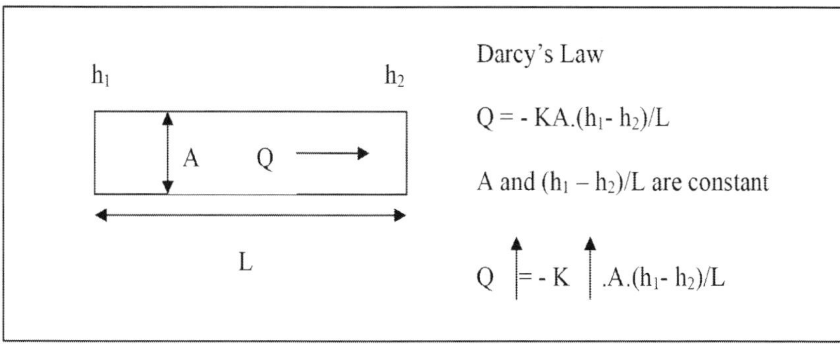


**Figure 1.2** Darcy's law.

**Source:** *https://www.researchgate.net/publication/264842134_Effects_of_Spatial_Variability_on_the_Effectiveness_of_Artificial_Aquifer_Recharge_in_the_Perth_Basin/figures?lo=1*

In the case of turbulent flow, there will be high-velocity rates. There will be an increase in the pressure gradient also. The pressure gradient may be higher than the flow rate. It is important to modify Darcy's equation accordingly. In the case of turbulent flow if we use Darcy's equation then we will get errors. In the case of steady-state flow the pressure, levels will remain constant along with the time. We can describe flow behavior of the fluids using steady-state flow. We can explain multiphase flow, radial flow of compressible fluids, radial flow of slightly compressible fluids, radial flow of incompressible fluids, linear flow of compressible fluids, linear flow of slightly compressible fluids and linear flow of incompressible fluids. (Gang Chen et al., 2016).

The pressure at the wellbore will decline if the well is opened. The preservations will move far away from the wellbore. The rate of this movement can be determined by Rock and fluid compressibility, fluid viscosity, porosity, and permeability. The radius of the pressure variation may increase with time continuously. This can be called the radius of Investigation. In the steady-state flow conditions, the quantity of fluid will be the same. We need to consider other variables as total compressibility, porosity, and time. The mathematical flow equations of the transient flow can be framed using three equations. They are continuity equation, transport equation and compressibility equation. We need to consider both initial and boundary conditions.

## 1.3 PERMEABILITY

The permeability (Figure 1.3) of the wellbore can be lowered due to Clay particles, cement slurry and mud filtrates that are derived from workover operations, completion, and drilling processes. This is called as wellbore damage. The region that is affected due to changed permeability can be called a skin zone. This zone can be thick and, in some cases, it can extend to several feet. The permeability that is far away from the well will always be different.

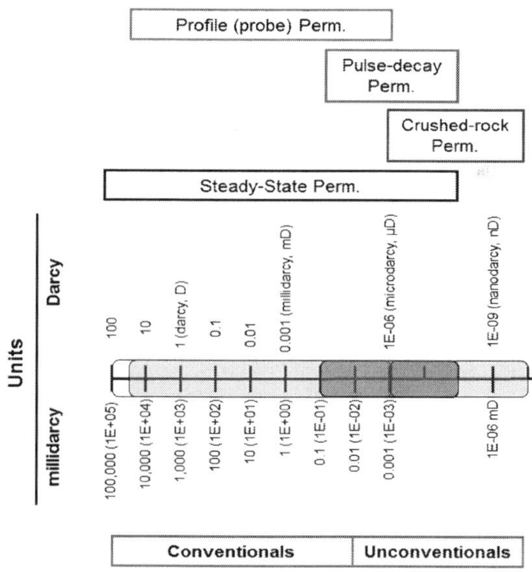


**Figure 1.3** Permeability.

**Source:** *https://www.researchgate.net/publication/287994894_Gas_transport_and_storage_capacity_in_shale_gas_reservoirs_-_A_review_Part_A_Transport_processes/figures?lo=1&utm_source=google&utm_medium=organic*

The reason behind this is that those zones are not affected by stimulation or drilling. The skin zone can alter pressure distribution in the vicinity of the wellbore. We need to consider the concept of superposition in the case of applying the transient flow equation. We can use this concept to know the pressure changes, boundary effects, rate changes and multiple well effects. Smoothing of time series graphs to pressure time is critical and a hard task to achieve. Interpolation, smoothing and signal filtering can be considered important in the smoothing process. The deviations of the semilow (Straight line) behavior will be based on the lateral enhancement of mobility, boundary, hydraulically fractured reservoirs, naturally fractured with device, stratified layers, interference, packer failure, phase separation, partial penetration, impermeable flow barriers and faults. Many theories describe unsteady state flow data. These theories are based on the radial flow of fluids in a Reservoir system that has uniform permeability, porosity, and thickness. The radial flow that is to be considered can be called an ideal radial flow. A minute deviation from this theoretical concept can drastically change the pressure measurements from the actual measurements of pressure. We can observe variations in the Well test responses. It is to be noted that the wellbore storage effect is important to describe the flow regime. We can notice the reservoir and well heterogeneities effect in case of pressure behavior responses. It may be due to the formation of fissures, hydraulic fracture, skin fracture and formation of multiple layers. The pressure responses may be radially infinite in their behaviors. This can also explain the homogeneous system. The boundary effects are also equally important (Robert., 1980).

## 1.4 FLOW REGIMES

There are several types of flow regimes that might have occurred during pressure variations (Figure 1.4). The identification of a model that fits the graphical analysis along with Computer-aided designs can be considered an important parameter. We can apply a pressure derivative approach as a diagnostic tool. In this approach, we can observe magnified pressure changes. The flow regimes can have obvious shapes that are characteristic of pressure plots. It differentiates several responses obtained from Reservoir models. Some of the responses that might be derived are infinite acting systems, impermeable boundaries, faults, constant pressure boundaries, closed boundary systems, etc. This approach can also identify diverse Reservoir behaviors and conditions that are hard to study using traditional approaches. We can get a high level of accuracy while estimating the test parameters. Some unlimited models canbe used in idealized conditions.

The validity of a well test and its interpretation can be based on certain factors. They are the applicability of the model and the accuracy of the field data. It is important to identify appropriate models to analyze data obtained from wells. The inner boundaries must be carefully studied after identification. We can assume five inner boundaries and respect to flow conditions in the vicinity of the wellbore. They are fracture, partial penetration, phase separation, skin, and storage of wellbore. The behavior of the reservoir can be important too. We need to collect middle time data. This can be obtained from infinite acting behavior. They can be classified into homogeneous and heterogeneous methods. We need to identify outer boundaries. This is essential because the model demands late time data. The outer boundaries may be classified into a constant pressure boundary and no-flow boundary. The early time data can be reliable and uses excellent information about the reservoir that is in the vicinity of the wellbore.

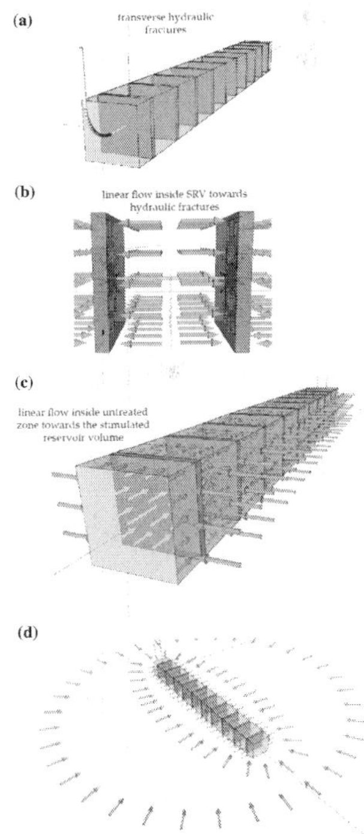


**Figure 1.4** Flow regimes.

**Source:** *https://www.researchgate.net/publication/288324030_Well_test_rate_transient_analysis_and_reservoir_simulation_for_characterizing_multi-fractured_unconventional_oil_and_gas_reservoirs/figures?lo=1*

The influx of water must be studied before the modeling process commences. The water-bearing rocks can be called an aquifer. These rocks can be present around the hydrocarbon reservoir. They may be larger than gas and oil reserves. When reservoir fluids are produced there will be a decrease in pressure and so the pressure differential will get generated around the aquifer. The aquifers react due to contact between hydrocarbon and water. There will be water encroachment because of hydrodynamics and formation recharge. This recharge can be due to surface waters located at outcrops. The pore volume of the reservoir can be larger than the pore volume of the aquifer. It is to be noted that the water expansion in the aquifer zones may be negligible if we take into consideration the total energy systems.

Mostly, the behavior of reservoirs can be based on volume. The influx of water may be ignored while calculating the behavior of the reservoir. The permeability of aquifers may be relatively low such that the pressure differential can be maintained. This will lead to encroachment of water into the reservoir. Most of the oil gas reservoirs can be produced by a water drive mechanism. This can be classified as a Natural water drive mechanism. In the case of artificial water drive mechanisms, there will be an injection of water into the reservoir formation. The production of hydrocarbons from the reservoir may decrease the pressure and hence the pressure of the aquifer can also decline. This can lead to water expansion in aquifers. There will be compression of rocks in the aquifer. The outcrop that is considered as a water bearing structure can be located at a higher position (Fikri et al., 2015).

The natural water drive can be classified based on the reservoir pressure maintenance degrees. They are an active water drive, partial water drive and limited water drive. If the rate of water influx is the same as the production rate of a reservoir, then we call it an active water drive. There will be a slow pressure decline in the active water drive reservoirs. The aquifers can be differentiated into finite and infinite. The finite aquifers can be called as bonded aquifers. Most geological formations are finite. The outer boundary can regulate the characteristics of aquifers. In the case of infinite systems, the effect of pressure variations at the aquifer oil boundary does not influence the outer part of the boundary. In the case of finite systems, the outer limits of the aquifer may be affected by the encroachment of the oil zone.

The Reservoir aquifer systems may be classified depending on the flow geometry. They are linear water drive, bottom water drive and edgewater drive. In the case of Edgewater drive, water can migrate to the flanks due to the production of hydrocarbons resulting in decreased pressure at the reservoir aquifer boundary. We can observe the radial flow and it will be vertical. Most of the bottom water drive can be observed in the reservoirs where there is a gentle dip and a large area. In this case, the contact between water and reservoir will be underground. We can observe radial flow in this case. In the case of linear water drive, then influx will be generated.

Limited information is available about the exploration of the development period of the reservoirs. We need to consider the previous performance data of the reservoirs. If we observe a continuous decline in the Reservoir pressure due to enhanced withdrawals can signify an influx of fluids. The water encroachment can be an indication of water production in the edgewells. It must be assumed that the early production of water might be due to fracture formation, coning and due to the formation of high permeability streaks. In some cases, the production of water may be due to leaks in the casing. The aquifer can be characterized by uniform thickness, constant permeability, uniform porosity, constant water compressibility and rock compressibility. In the process of estimating the cumulative water influx, it is essential to take into consideration the influx intervals from the beginning (Neil., 2008).

The dam projects may be classified into storage projects and run- of-river projects. Most of the hydropower generation projects can be classified under run- of- river projects. These projects have minimal storage volumes that are active. They will be having large storage volumes that are dead. It is important to increase head to improve storage according to the projected demands. The storage projects can have large storage volumes that are active. They will be having Limited storage volumes that are dead. The active storage volumes may contain the water needed for water supply and irrigation. We can use these projects to reduce floods or to manage floods. We can use storage projects to produce hydropower. In some cases, the requirement of heat will differ. The extent of the head can be important for the production of power.

The objectives of sediment Management can be based on the projects which we choose. In the case of run -of- river projects, the primary aim of sediment management is to enhance the operational efficiency of the project. If there is improper sediment clearance, then there will be abrasion in the turbines and hence the cooling water intake will be damaged. The electromagnetic equipment will get damaged due to sediments. This will drastically enhance the maintenance and operational expenditures. The power generated from these projects will also be reduced. We need to consider Reservoir storage in terms of sustainable development. This can be important in supplying renewable energy and water production. This space needed for storage in the reservoir must be considered as a Natural Resource. In the presence of dams, the storage space must be considered as an important natural asset. The storage space must be assumed as an exhaustible resource or a renewable resource. This can be determined while designing and operating the reservoir and dam. If we put the water into the reservoir to carry sediments and reduce the storage capacity of the reservoir, then we need to consider it as an exhaustible resource. If we can reduce the storage losses due to sedimentation then we can assume it as a renewable resource.

The sedimentation of reservoirs can be due to the inflow of river water that can deposit increased sediments in the dam upstream. The reason behind it is that the speed of the inflow river will be drastically reduced and hence the rate of sedimentation will increase within no time. The water may not have enough force to participate in sediment transportation.

Chapter

2

# River Flow

## CONTENTS

2.1 Introduction .......... 14
2.2 Hydrological Conditions .......... 16
2.3 Fluvial Geomorphology .......... 17
2.4 Rills .......... 19
2.5 Streams .......... 22
2.6 Climate Associations .......... 32

## 2.1 INTRODUCTION

Most of the large ecosystems of the world are being stressed. Due to the increase in the population growth, there is continuous pressure on the water supplies, land resources and hence destroying the native species. Most of the cultural resources are being lost. Several asset uncertainties are to be evaluated in the present and future (Hillman et al., 2016). Several societal challenges are to be addressed through innovative policies and governance. River management plays a major role in assisting us to develop certain alternatives that can be easy for us to understand the quality of life and ecosystem sustainability (Jens et al., 2019). You can identify the driving forces that are damaging the local environment. Due to population pressure, Several monitoring techniques have recently emerged. To help us understand the course corrections in the adaptive strategies, ecological studies have become an essential part. In most of the research environments and practices, the geomorphological features will dictate the course of rivers and they act as indicators. In case of floods the aquatic habitats can be sensitive (Figure 2.1).

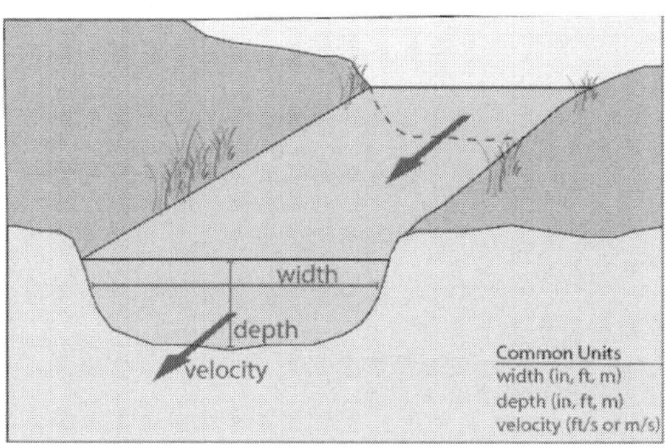


**Figure 2.1** River Flow.

*Source: https://www.e-education.psu.edu/earth111/node/864*

Several challenges can dwindle the balance between the use of river resources and human encroachments. It is essential to go for sustainable ecosystem management. The physical system, wherein humans can survive can be dictated by the local river systems. Several ecosystems are being damaged due to population explosion and unsustainable use of water. The invasive vegetation is dominant than most of the native species are at stake. It is essential to go for experimentation of the strategies that are complex and dynamic. As far as the adaptive management framework goes, it is essential to explore the key species that are damaging the river environment (Firat., 2007).

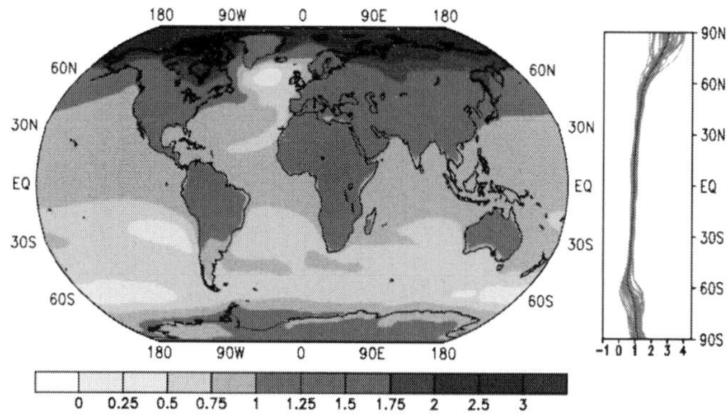


**Figure 2.2** Climate change patterns.

**Source:** https://www.researchgate.net/publication/332264381_The_Polar_Amplification_Model_Intercomparison_Project_PAMIP_contribution_to_CMIP6_Investigating_the_causes_and_consequences_of_polar_amplification/figures?lo=1

It is expected that the climate change patterns (Figure 2.2) can drastically change the course of watersheds and the local areas and it can be upskilled to the global areas to get a change in the watershed patterns. There will be a disturbance in the sediment yields and flow distribution, at least in the fluid system. It is essential to use tools that give accurate measurements so that we can frame certain policies to nullify the problems created due to the population explosion. The complexity of our management can be dependent on the cross-sectional area, banks and riverbeds. Each configuration of these components can drastically affect the health of the river. There are several problems with dams and their associated channels. We need to go for sustainable river restoration processes and that includes the characterization of forms and magnitudes of physical complexity that are happening in the river environment. We need to understand the physical complexities, as and when required and based on the problems to be faced. It stays to the degree that the physical complexity can be at least restored to a minimal level, at first. These demand educating the stakeholders to improve the fiscal complexities in the reversal of damage. We need to bridge the gaps that are stopping the local communities from damaging sustainable ecosystem management.

Most of the structures of the river can be based on the geomorphic context and the history of the river structure. We need to understand the diversity and the habitat abundance at the first opportunity if we want to understand the river itself. Most of the rivers will have spatial heterogeneity, especially with the channel boundaries. We need to understand the spatial heterogeneity substrate, grain size, depth of water and retention times. There will be a lot of heterogeneity in the flow hydraulics substrate and channel geometry. This will drastically enhance the diversity of habitats for aquatic organisms and riparian cultures. The physical complexities can regulate the resilience of the rivers and resistant management practices. The physical complexity can be completely changed due to human-induced disturbances in the river ecosystems (Alberto., 2019).

Most of the river ecosystems will resist change and there will be floods in rare scenarios. Naturally, the resilience of the river ecosystem can be defended by the tendency of the river to get back to its pre-disturbance conditions mostly encounter. Different spatial complexity can be present in the form of secondary channels and floodplains. The floodplains play a critical role during river flow and its management. These secondary channels can buffer the droughts and floods acted upon by external forces, nutrients sediments and water retention capacity can form an important base in defining the physical complexity of the rivers (Schodro., 2007).

## 2.2 HYDROLOGICAL CONDITIONS

The rivers can be considered as an integral part of the hydrological cycle (Figure 2.3) that occurs inside and outside of Earth. The rivers are prone to several pressures and modulations from the landscape. The anthropogenic pressure can also add to this and hence most of the river flow dynamics will be subjected to both natural and anthropogenic pressures. Most of the surface of Earth will be covered with water. So, the hydrological cycle will be altered predominantly by the water bodies than the landscape dynamics. It is to be known that oceans occupy at least 71% of the surface of Earth. The glacial ice will cover at least 10% of the land surface available. Just 0.58% of the surface is covered with rivers (Bear., 1981). The groundwater will also support the terrestrial life as the surface water. Through condensation, advection, sublimation and evapotranspiration, water will be shifted or transported into the atmosphere. This water can move over long distances and so the temporal and spatial analysis of the water collected from the terrestrial sources will be critical. There will be a lateral shift of the water over the atmosphere. Then, the water can get deposited over other landforms or oceans. Water that is deposited on the land surfaces will reach oceans through diverse atmospheric processes, lateral groundwater flow and runoff. Rivers can be classified as natural features mostly involved in feeding oceans whilst passing through several undulations of the landscape. The transport of water will involve erosion and the solution of the materials present on the surface of the earth. Most of these contents will be transported into the oceans or any massive bodies in the vicinity. The rivers can transport water and sediments that constitute the biogeochemical contents. These contents are essential for the living organisms that survive on the surface of the earth and inside it.

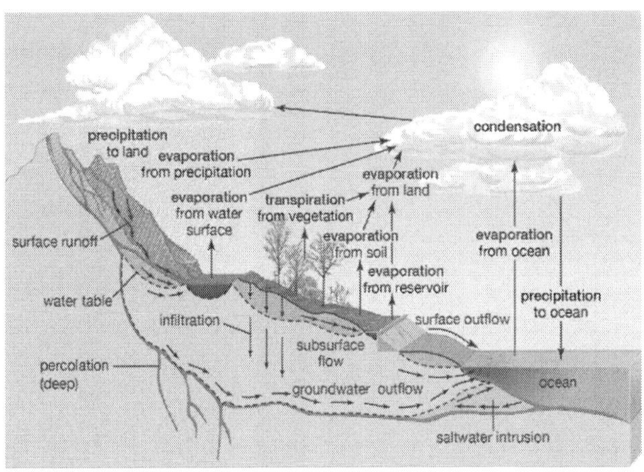


**Figure 2.3** Hydrological Cycle.

**Source:** *https://cdn.britannica.com/89/62689-050-BD53B2F5/water-hydrologic-cycle-land-surface-atmosphere-ocean.jpg*

## 2.3 FLUVIAL GEOMORPHOLOGY

The fluvial geomorphology (Figure 2.4) is important whilst investigating the river dynamics and its course (Stephen et al., 2005). This branch will be dealing with the river dynamics mostly and it also describes several other features of the river movements. The abductive and deductive reasoning can be classified as important types of scientific reasoning that can be studied in fluvial geomorphology. These reasoning types can be explained by theory.

Most of the diffusive processes can be dependent on the slope and they may be devoid of eroded sediment and transport of earth materials. The soil creep can be distinguished by the slower movement of the soils according to the slope and this can be affected by gravity. Most of the rainfall events can alter the downslope displacement of these particles. This is because of the raindrops that might fall on the surface of the earth. The sediment flux can be enhanced with increased slope. Mostly, the sediment transport will be from high slope conditions to lower slope conditions. Most of the depressions in the landscape or over the landscape will be filled with the sediment debris that is brought about by the events that occur upstream. The elevated surfaces will be eroded and this will result in a permanent change of the landscape. We can expect topographic irregularities and relief decrements due to the action of water. Most of the advection processes will be based on the hydrodynamic properties and this is directly based on the flowing water dynamics. The advective transport can eliminate the sediments from any location. We can frame a model based on the channel initiation and this can be due to seepage erosion. Most of the seepage erosion can be due to the upliftment of the rocks. The rock mass that is being uplifted may exhibit uniform permeability. There will be hydraulic gradients that can assist in the flow of groundwater. The diverse rock characteristics and chemical weathering rates can result in the spatial variation in terms of permeability. There will be local perturbation. This will yield concentrated flow of groundwater (Kondolf et al., 2005).

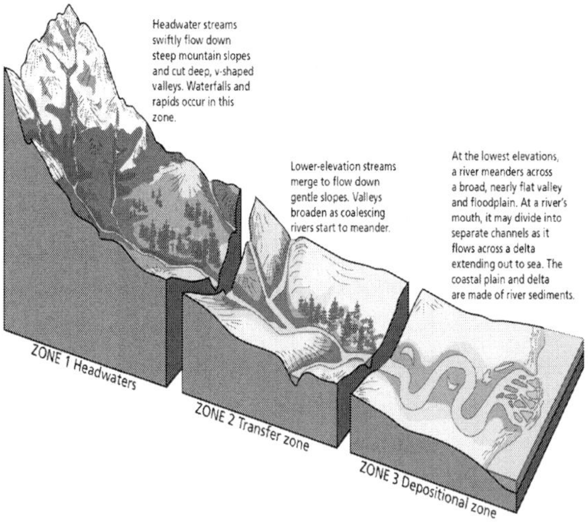


**Figure 2.4** Fluvial Geomorphology.

**Source:** *https://www.nps.gov/subjects/geology/fluvial-landforms.htm*

## 2.4 RILLS

Rills (Figure 2.5) must be studied whilst understanding the river flow characteristics. Most of the rills will be formed on the hill slopes that are soil mantled. The gradients that can support the formation of rills will be in the range of 2 to $12^0$. There will be a negative feedback system in the rills. This will lead to either eradication or stabilization of channels and this will progress according to time. There will be flow channelization in the soil particles that are coarse and concentrated.

**Figure 2.5** Rills.

**Source:** https://www.researchgate.net/publication/249009078_GIS-based_soil_erosion_modeling/figures?lo=1

There will be inhibition of incision. The incision can get impeded due to clay hardpans. They can be called as resistant subsurface layers. There will be a reduced volume flow rate due to the infiltration of runoff into the rill of the bed. All these processes can reduce erosion. The incision of the rills can be counteracted due to the rain splash process and this can be considered as an effective process. Some of the rills can produce gullies. These are classified as erosional features that can be more massive than the rills. We can classify these gullies as enlarged rills. The erosional features might occur in the hill slopes that are steeper. The size and other parameters can be different and this will damage the tillage operations. The gullies and rills can share similar morphology and their size differs. The transitional forms can yield distinguishable features between gullies and rills. The formation of gullies can be an indication of the high rate of erosion that might occur on the hill slopes. This can be due to anthropogenic pressures and natural pressures (Robert et al., 2010).

Climate change can moderate the vegetation cover dynamics and precipitation intensities also vary. This will relatively increase the erodibility of the landscapes. Urbanization, deforestation, farming, and other human activities can reduce the soil strength and hence its resistance to erosion will be affected. There will be an enhanced rate of runoff. The gully formation can be considered as an important sign of land degradation. The formation or initiation of gullies can reflect positive feedback mechanisms. The flow concentration can be within the limits of the incipient channel. This concentration can go beyond the material resistance on the bed. The gullies can be erosional features on the hillslope. We may not consider them as a part of the incipient stream channels. Some of the gullies can be larger and they can be categorized into permanent entities. The permanent and large gullies can be discrete. They may not get connected with the downslope with either streams or rivers. Most of the gully networks will be restricted to the hill slopes with poor networks (Pierre et al., 2005).

The total sediment flux can be important while studying the river path dynamics. There are several factors that can affect sediment mobilization on the hill slopes and the channels of streams. Some of the eroded materials can also affect the river flow dynamics. The flow capacity, effectiveness and amount of flow can affect the spatial and temporal behavior of transport and mobilization regimes. If there is the mobilization of the materials, we can expect spatial continuity in the transportation process. These flow variations can also generate the temporal variability across several spatial dimensions. In most of the studies, vegetation and climate were discussed as independent and autonomous entities that might affect the sediment fluxes. Mostly climate can be studied in terms of temperature and annual precipitation patterns. It is to be known that once we get the annual precipitation means, we can estimate the probable runoff. This will also give us certain ideas about channel erosion and increased hillslope per se. It is affirmative to assume that the sediment flux is higher in case of wet conditions than the dry conditions. It is to be understood that temperature fluctuations are important in determining the sediment fluxes. This can be based on the landscape orientation and its susceptibility to erosion (Herve et al., 2005). The hot and cold conditions can also diminish the vegetation growth rates that can be attributed to erosion and runoff. Permafrost can keep the soils tight and compacted so that the erodibility will be diminished. In snowy conditions, precipitation might land on the surface of the earth in the form of snow. This may minimally contribute to the surface runoff. Vegetation can also control the erosion rates at the hillslope regions. This can be attributed to the lowered rates of runoff. The rainfall that is being precipitated over the land surface will be intercepted by vegetation. Because of vegetation, the shear strength of the weathered material and soils will be enhanced. This will drastically enhance the hydraulic resistance of the soils and keep them saturated as time progresses (Marwan et al., 2005).

The environments that are specified to support thick vegetation and so the erodibility levels can be minimal than the drylands. In the case of drylands, the weathered materials and soils can be exposed to the runoff directly and hence loss of sediments is evident. So, it should be advocated that enhanced vegetation levels can lower the sediment flux rates. The topography of the basin can alter the sediment flux. This can be achieved by changing the hillslope gradients. The soil erodibility and rock erosion or weathering can affect the sediment flux rates. The drainage basins can be generated in the highly eroded rocks and weak rocks. The prominent rocks that get eroded are mudstones and shales. This will yield more sediments than resistant rocks. Schist and quartzite can be called resistant rocks. The weathering and jointing process can also affect erodibility and rock strength. There will be increased sediment flux if the erodibility rates get increased. The soils that are resistant to erosion might generate limited sediments than the erosion-prone soils (Reid et al., 2005).

It is difficult to develop a relationship between drainage area and sediment yields. The drainage area and sediment yields can be scale-dependent. The variations in the depositional and erosional processes within the channels and over the hillslopes can be considered. The soil erosion can be considered as an initial or primary process that can add to mass flux in the drainage basin. The eroded soils can be transported across the downslope and this can be mainly due to gravity. These eroded soils will be subjected to diffusive and advective processes. Flow dynamics can be an important study area that is to be dealt with in detail. The rivers can be classified as channeled flows of water and these channels can reach the outlet once the sediment load gets exceeded. The study about the fluid flow in pipes and channels can be called hydraulics. Fluid mechanics can be considered as a fundamental core of hydraulics. We need to consider the flow dimensionality and classification to understand the complexity of the river flows (Hooke et al., 1995). This can be assumed as a qualitative framework. If the flow becomes unsteady, then we can assume that it is complex. The unidirectional flows, uniform and steady flows can be classified as simple. The unsteady flows can be demarcated with the three-dimensional fluid motion and often characterized as complex. The flowing water can exhibit turbulent and laminar behavior. To understand the river flow dynamics, we need to understand these behaviors. On application of shear stress or in response to that, the laminar flow can be characterized by thin layers of flowing fluids that can be parallel to each other and the velocities will be different according to the depths. Mostly, the layers that move fast can be found in the surface zones. The surface fluid layers will be entangled by the air that is being overlaid on the layers. It is worth noting that the water molecules at the base or that are attached to the bed exhibit limited or no movement. Between surface and bed, the layers will be subjected to varying speeds. The increase of layer velocity can be from the bed to the surface. In the case of laminar flow, viscosity can be considered. This can determine the friction between the layers and hence can be studied to know the magnitude of the internal resistance of the fluid at movement (Ian et al., 2019).

## 2.5 STREAMS

In the streams (Figure 2.6) and rivers, there will be friction that can exist along the boundary of the channel that resists the water flow downstream due to gravity. The mean velocity can be determined by friction and gravity in the case of steady uniform flow conditions.

**Figure 2.6** Streams.

**Source:** *https://www.usgs.gov/mission-areas/water-resources/science/stream-ecology?qt-science_center_objects=0#qt-science_center_objects*

There will be a balance between the resisting forces of friction and driving forces due to gravity (Roey et al., 2014). The shear stress on the bed can affect the frictional resistance of the river flow at various spatial scales. This will be in the limits of the channel length over the energy slope and bed. This range can be in the limits of planform scales, bar-unit, bar-element, bedform and grain. The morphological contents at varied spatial and temporal scales can affect the shear stress of the bed and reach-scale resistance. The grain resistance can be considered as an important feature in the beds at the rivers that are predominantly covered by boulders, cobbles, and gravel. The resistance offered by the bedform can affect the river flow in the river beds. The height of the ripple bedform can be between 0.01 to 0.04m. The height of the dune bedform can be between 0.1 to 10m. The height of the antidunes (Figure 2.7) can be in the range of 0.03 to 1m (David., 1999).

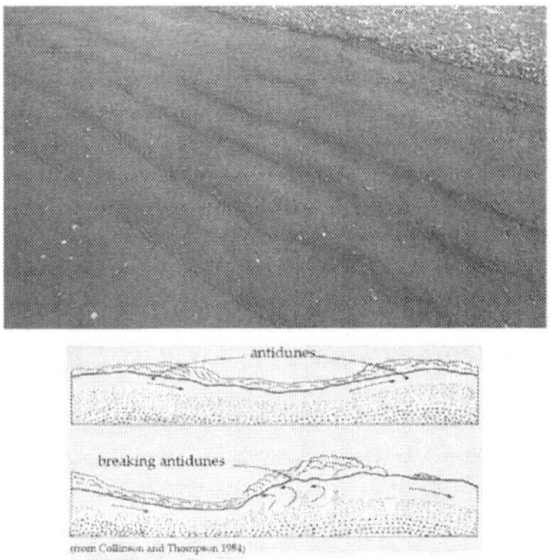


**Figure 2.7** Antidunes.

**Source:** *http://www.columbia.edu/dlc/cup/ricci/ricci13pl37.html*

The ripples (Figure 2.8) are nothing but a tiny wave-like force that occurs in the crests. They can occur perpendicular to the direction of the flow. We can observe asymmetrical longitudinal profiles. It will have a steep lee side and a gentle stoss side. The sediments can be transported downstream. This form comprises coarse sand. The flat beds can exhibit limited to negligible irregularities. Most of the sediments can get migrated in the form of bedload sheets. This form is made up of fine gravel and consists of coarse sand.

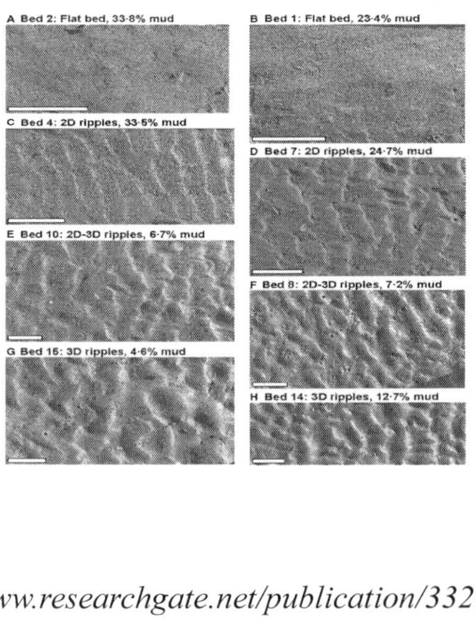


**Figure 2.8** Ripples.

**Source:** *https://www.researchgate.net/publication/332193981_Integrating_field_and_laboratory_approaches_for_ripple_development_in_mixed_sand-clay-EPS/figures?lo=1*

The dunes (Figure 2.9) can be differentiated by large wave-like forms. They can form in crests that can be perpendicular to the flow direction. In some cases, they can form at high angles against the flow direction (Daniel et al., 2014). They can be out of phase with the surface topography of water. There will be an asymmetrical longitudinal profile. We can observe a steep lee side and gentle stoss side in this case. The transport of sediment can be downstream. The upper stage plane bed is another bedform that is to be studied. They can emerge as dunes washouts.

**Figure 2.9** Dunes.

**Source:** https://en.wikipedia.org/wiki/Dune#/media/File:Dyyni_Yyteriss%C3%A4.jpg

Most of the sediments can be migrated or transported in the form of sheets. We can observe bed waves that are of low relief. They can represent the dunes that are being degraded by eradication. Most of the dunes cannot grow in height. The antidunes can exhibit wave-like forms that have low-amplitude and they can be large in certain cases. These forms can occur perpendicular to the flow direction and they can also form at high angles to the direction of the flow (Bruce et al., 2003). They can exist in phases with the surface topography of water. We can observe symmetrical longitudinal profiles. Gradually, they may become asymmetrical. These waveforms can occur on the surface. They can be stationary and rounded. In some cases, we can observe the breaking of the formation. These bedforms can easily get transported downstream, upstream and in some cases, they can be stationary. The pools and chutes are other dominant forms that can be observed in the rivers. They can be in the form of elongated mounds. These sediments can be in the form of steep chutes. We can notice supercritical flow in the troughs and pools of the subcritical flow. There will be deposition in the pools and we can also expect the erosion of the chute slopes. This will result in the migration of sediments to pools and chutes upstream. In this case, we can expect a Froude number that ranges between 1 and 2. There will be local deposition of the sand particles that can yield pebble clusters (Alexander et al., 2018). As the Froude number increases, there will be varying flow resistance that results from the periodical changes in the bedforms. Based on the gradual increments in the Froude number, we can observe immobile plane beds initially. Then, we can expect lower plane beds and ripples. We can observe dunes and washed-out dunes in the process of transition. We can also observe upper plane beds and antidunes based on the gradual increase in the Froude number. The boundary layer in case of turbulent flow may contain various components. The roughness and smoothness can be considered important aspects of the particle motion on the bed (Assani et al., 2005).

The transport of the sediments in the rivers can be considered as a basic process that links the flow dynamics with the river systems. In the case of alluvial rivers, most of the channels will be carved into the materials that might have gradually deposited by the river. This deposition can result in the formation of the flood plains. The deposition and erosion processes can be linked with the transport of the sediments. These processes can determine the shape of the river form. They can also define the floodplains and alluvial river channels. In the case of rivers with bedrock, the sediments that are being transported can play a vital role in changing the channel beds or we can consider it as abrading of the bed channels. This can result in erosion and it will tend to increase with time. There is an important link between river morphodynamics and sediment transport. It is important to understand the processes underlying the deposition and erosion of the sediments. The material flux generated from the rivers may contain solid loads and dissolved loads. This will be based on the area which is being bisected with the river bodies. The chemical weathering of soils and rocks can bring about solutes. These solutes will be transported into the rivers and this can be mostly due to surface runoff or we can call it overland flow. In some cases, the flow can be through soils and this can be called as through flow. In some cases, we can expect groundwater flow (Figure 2.10).

The rivers might contain mineral and rock particles and this can be called clastic materials. We call it solid load to describe the materials that are being deposited in the river beds. The solid loads can be used whilst investigating water quality parameters. The clastic materials can be migrated via rivers. The actual source of sediments that are being transported can be located exterior to the river channel or inside the channels. The order of size of the sediments can be clay<silt<sand<pebbles<cobbles<boulders. The pebbles, cobbles and boulders can be classified as gravel (Yoram et al., 2005). The out-of-channel source can be called a watershed. It involves both soil erosion and mass movement.

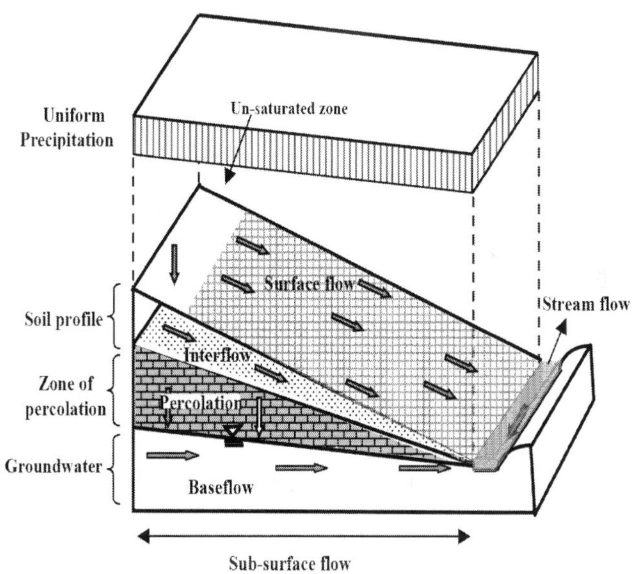


**Figure 2.10** Groundwater flow.

**Source:** https://www.researchgate.net/publication/228491737_ANN_Modeling_for_Estimation_of_Surface_and_Subsurface_Flows_Based_on_Watershed_Geomorphology/figures?lo=1

The soil erosion (Figure 2.11) can be in the form of gully erosion (Figure 2.12), rill erosion and sheet wash. We can observe clay, silt, and sand in this case. In the case of mass movement, we can expect flows, falls, and slides. We can observe clay, silt, sand, and gravel in the sediments formed due to mass movement. The In-channel sources can comprise bank erosion and bed erosion. The upper cohesive bank can constitute clay and silt particles. The lower non-cohesive bank erosion can result in sand and gravel. The bed erosion can yield thalweg and bars and the sediments formed due to bed erosion can constitute sand and gravel. The transport mode can be classified into suspended load and bed material load. The suspended load can be differentiated into wash load and suspended load (Rukavina., 1996). The bed material load can be demarcated into bed material and bedload. The saltation process can be classified as bedload.

**Figure 2.11** Soil erosion.

**Source:** *http://www.omafra.gov.on.ca/english/engineer/facts/12-053.htm*

**Figure 2.12** Gully erosion.

**Source:** *https://www.sciencedirect.com/science/article/abs/pii/S016955X17301277*

In the case of wash load, we can expect clay and silt. In the case of suspended loads, we can observe sand and portions of gravel too. The bed load comprises sand and gravel and clay is rarely seen. The mass movement and soil erosion can be labeled as an out-of-channel process that can help in the sediment transport to reach rivers. Soil erosion can be expected in diverse environmental setups. In this case, a mass movement is prevalent especially when there is steep terrain. We can observe this type of erosion in mountain terrains. The soils can constitute sand, silt and gravel and these particles will be supplied to the rivers in diverse proportions. Though several coarse particles are present, soil erosion may not be able to transport massive particles into the river streams. Mostly, the particles that are larger than the pebbles will not be transported usually. The mass movement process can easily transport particles of varied sizes like boulders and cobbles into rivers (Scott et al., 1980).

In the case of In-channel sources, most of the sediments comprise channel bed and channel banks. The alluvial rivers can have channel banks. They constitute a cohesive upper part dominated by fine sand, clay and silt. In the case of a lower non-cohesive portion, there will be particles that are the size of boulders and sand. The channel bank erosion can result in coarse and fine particles that might enter rivers. The channel bed texture can be based on the size of the particles that will be deposited into the rivers. The coarser particles can be present in the alluvial river system and mostly they reside in the channel beds. Some of the important features of the bed are thalweg, bars and local sediment deposition. The particles that are present on the bed can be transported along with the flow in the channel. This can evolve gradually as an important source of transport. The transport of sediments in the rivers can be divided based on the method of transport. The classification can be based on the material that is to be transported. The sediments can be transported in the rivers through hopping, sliding and rolling or through suspension in the water column (Forbes., 1976).

Certain variables are to be considered whilst studying the dynamics of the sediments. Some of them are discharge, peak discharge, duration of runoff, total water volume, rate of discharge variation, sediment storage within the channel, pathways that connect the pathways of the runoff according to the sediments. We need to consider the spatial extent of the runoff. We need to measure the time taken between mobilizing events. It is essential to know the antecedent soil moisture. We need to take into consideration antecedent runoff and antecedent precipitation. We need to know the amount of precipitation and duration of the rainfall events. We need to know the intensity of precipitation too. The seasonal variations can affect sediment concentration. The land cover characteristics and precipitation can also affect the sedimentation process. We need to study the storm types along with the precipitation intensities. It is essential to know the patterns of the precipitation events. There will be variations in the vegetation characteristics also. The vegetation dynamics can be affected by the local climatic conditions and this can deeply affect farming activities. We need to consider the seasonal differences as regulating factors of sediment concentrations. The sediment concentration and discharge can be studied in detail in terms of seasonal variations. In the case of event scale, we can expect complicated sediment dynamics. Several factors should be considered while explaining the sediment systems.

The mechanics of the sediment transport (Figure 2.13) and flow can be via fundamental mechanisms. Some of the mechanisms are the interaction process and morphology of the river systems under study. The flow of rivers can be estimated by fluctuating hydraulic conditions, rate of discharge and discharge dynamics over time. The changing hydraulic conditions can bring about change in the transport of sediments. This will affect the channel morphology.

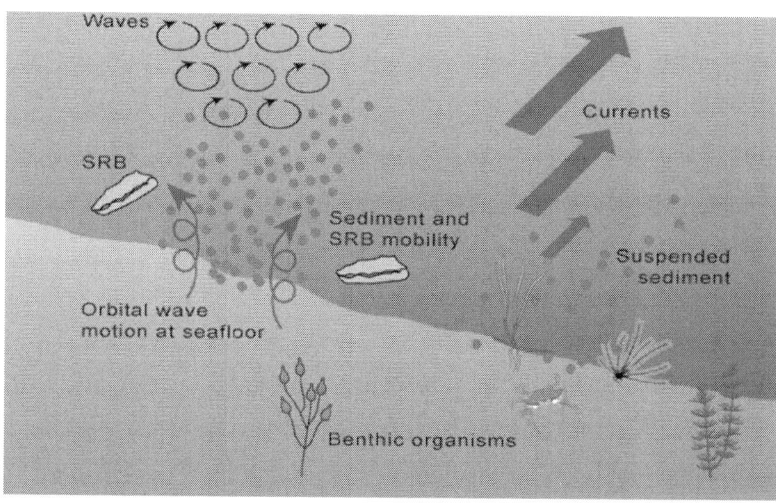


**Figure 2.13** Sediment transport.

**Source:** https://pubs.usgs.gov/of/2012/1234/figure7.html

We need to study the river dynamics in connection with the spatial-temporal variability of the discharge by rivers. The flow of rivers can be regulated by the land surface tier of the hydrological cycle. Water can be supplied to the rivers through runoff preferably from the watershed. Most of the precipitation that falls over the hillslopes can get infiltrated into the soils. This infiltration can be based on the permeability of the soils. This can be studied in terms of hydraulic conductivity. It is nothing but the movement of the water via soil particles. We need to consider the antecedent conditions pertaining to moisture too. The soils that exhibit high hydraulic conductivity and high permeability rates can exhibit high rates of infiltration. The moisture can gradually close the pore spaces. Some of the moisture that filled the pore spaces can enter the deeper zones. The water reaching the subsurface can form a wetting front. Once the pore spaces get filled, the rate of infiltration at the surface will start to decline. It is to be noted that all the water deposited due to precipitation may not reach the interior soil layers. In case of saturation of the soils at the surface, there will not be further infiltration.

The intensity of the rainfall events can be intercepted by vegetation. This will relatively permit excess infiltration rates. Therefore, we can expect more infiltration over land. The surface stormflow can result due to excess infiltration overland and overland flow. This problem will get enhanced in case of a storm surge post rainfall event. Water can get transported down slope and lateral modes and this will be based on the soil features. There will be several pressure gradients that may exist in the local areas. The bankfull channel can be considered as an essential element that can be classified as a morphological element especially in the case of alluvial rivers. We can understand this concept with the aid of formative flows that might have shaped it entirely. This channel can accurately describe the slope, depth and width of the flow conditions. These channels can affect morphological features. The rivers and streams that may flow via bedrock can have limited alluvial soils. They may not exhibit bankfull morphology completely. It is hard to identify the bankfull stage in the case of braided rivers. This can be attributed to the formation of the multiple channels that cross floodplains. This can represent complex topography. The bank heights and alluvial rivers can vary on either side of the channels. It can be prominent in the case of rivers that are being meandered. We need to know that there will be a lower inner bank when compared with the outer bank.

The effective stresses caused due to geomorphic features can change the morphodynamics of rivers. The extent to which the river system gets altered by hydrological events followed by recovery can be based on the relief of the land, drainage area, the thickness of soil, watershed characteristics, vegetation properties, geology and climate. There will be non-equilibrium morphodynamics and transient forms in semiarid and arid regions. This can be observed in the case of ephemeral rivers. We can observe equilibrium morphodynamics and characteristic forms in the case of perennial rivers. This can occur in humid-temperate conditions. The rivers in the arid regions can also present morphological behavior that can be transient. Mostly, there will be no adjustment due to sediment transport regimes and hydrological conditions. The morphology of the rivers can represent the formative flows that have shaped the rivers for decades if not centuries. In rare cases, we can expect both nonequilibrium and equilibrium forms in arid conditions (Tooth et al., 2000). The spatial scale can be considered as an effective geomorphological feature that must be considered (Fryirs, 2017). There are certain confined reaches that can produce greater stream power which can yield severe erosional changes. The unconfined reaches can be loaded with sedimentary deposition.

## 2.6 CLIMATE ASSOCIATIONS

The climate conditions at a local scale can play a vital role in regulating the effectiveness of floods, especially in the lowland ecosystems. In the case of lowland meandering, the impact of changing morphology can be limited or negligible. This can be due to fine bed material, vegetated channel banks, limited hydrological fluctuations and minimal stream power. These channel banks consist of clay and silt in excess amounts and hence they can be resistant to the erosion forces. The rivers in the semi-arid and arid zones can be of dryland fluvial type. These can be called flood-dominated systems. Because of the extreme dryness, there will be increased hydrological stability. There will be substantial variation spatially and temporally across the dryland ecosystems. There will be an abnormal widening of the river flow due to the occurrence of floods. The structure of the river can be classified into three components that can be distinguishable from superficial properties. The important properties are elevation change, planform features of rivers and geometry of the channels. The amount of water can vary in the river as time progresses. The hydrological variability can either get decreased or increased. Spatial variability of the hydrological parameters can be studied to know the river flow course (Joseph., 2018).

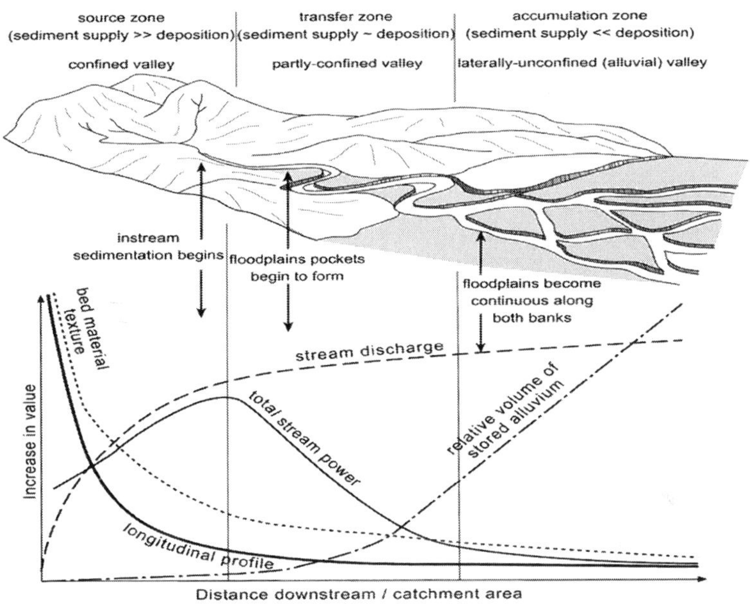


**Figure 2.14** Channel geometry.

**Source:** *https://link.springer.com/chapter/10.1007/978-3-319-73250-3_3*

The analysis of the channel geometry (Figure 2.14) can be useful in the quantitative estimation of the flow. With the data obtained from this analysis, we can know the dimensions and limits of the stable channels. Most of the channels that are being passed onto the bedrocks can represent channel geometry relationships downstream. These relationships can be almost like the alluvial channels. The river pattern that can be viewed from the above or aerially can be classified an important characteristic of the river morphology. The aerially viewed characteristic can give us an idea about the river morphology. The aerial view can give us an idea about the shape of the river and the planform of the river. We can yield a two-dimensional horizontal structure of the river. We can get information about the river morphology through an image or a map. The number of channels can be important to classify the rivers broadly. We can differentiate rivers into single-channel and multiple channel rivers. We can differentiate braided rivers in the case of multiple channel rivers. We can divide the channels temporarily. We can define the deposits that are not of vegetation as bars. These rivers can be called anabranching rivers and bars. The local environmental characteristics can affect the channel planform. Some of the essential environmental factors are bank erodibility, sediment supply, bed-material caliber, the slope of the valley and discharge. These factors can affect the inputs being supplied to the rivers. The important inputs are sediment supply and discharge (White., 2018).

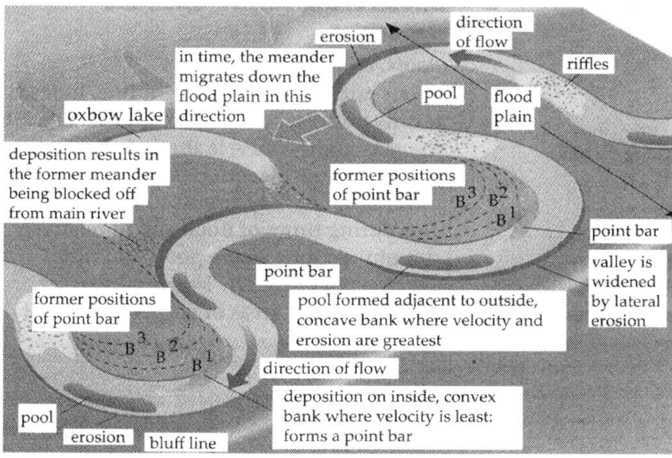


**Figure 2.15** Meandering of rivers.

**Source:** https://link.springer.com/chapter/10.1007/978-3-319-73250-3_3

The meandering of rivers (Figure 2.15) can be considered as an important aspect that is to be studied. The single thread channels can wind across the landscapes. There will be the formation of curves that can be reversed and they can be called as meanders. This is observed in most of the river systems. The meanders can be based on the river type. We need to understand the meander formation process in detail and this will help us understand the river dynamics. The meander dynamics can be associated with the change and development of the floodplains. There several mechanistic theories that can describe the meanders and there will be the formation of curves. The meandering streams can be produced on the surface of the glaciers. The channels at the glaciers will be carved (Helgi et al., 1989).

Chapter

3

# Environmental Management of River Basin

## CONTENTS

3.1 Introduction .......... 36
3.2 Flooding .......... 37
3.3 GCMs .......... 45

## 3.1 INTRODUCTION

The rivers can be classified as an important source that mankind can afford since the start of civilization. The rivers can play a vital role in providing water for industries, agricultural activities, and domestic needs. They can also generate nutrients and sediments that can support the natural ecosystems. Most of the landscapes can be classified according to the river basins which can be mappable. They can be mapped at various scales. We can divide river basins into lower, middle and upper basins. This classification can be based on geomorphic and hydrological characteristics. It is worth noting that no two rivers may not share the same response if acted upon by natural processes (Roger et al., 2005). Every river basin is unique in its way to react against the natural processes. Any conservation and developmental activity must be systematically planned to produce limited damage or negligible damage to the river ecosystems. The river flow may transfer fertile soil and water from the areas that are destined as catchment areas. They can distribute the sediments to the lower reaches. The sediments and nutrients can aid in the cultivation and formation of settlements. The water, sediment transport and channels can occur in the river basin.

There will be a unified system that can connect all these features to maintain environmental equilibrium (Popovici., 2014). This equilibrium can be affected by slope, lithology and climate, etc. If the river systems are not disturbed, they can maintain native dynamism to manage equilibrium conditions. Any disturbance in the river system can be due to erosion and flooding events. Desertification can also bring about changes in the river ecosystems. All the disturbances in the river basin systems can damage the commercial activities, agricultural production and land that is inhabited. This minor disturbance can change the availability of nutrients, especially in the deltaic regions. This can result in the expansion of exotic species mainly in the coastal regions. This will drastically bring down the natural products that are being produced and hence people dependent on this production may get affected. The dynamics of land use can impact the river catchments continuously and this will bring about a negative impact on the river basin structure. The urban sprawl can damage regional and local hydrology. The quality of water will be damaged. Due to urban development, the non-point sources of pollution will increase and hence the local natural resources will be damaged (Naoki et al., 2003). Catchment erosion can be related to anthropogenic activities. Most of the anthropogenic activities will lead to nutrient loads, diminished dissolved oxygen levels, acidity, turbidity, ground and surface water pollution. There will be increased sedimentation levels due to human-induced activities. Geohazards can also affect the structure and quality of rivers. Quicksand, cyclones, storms, tsunami, sedimentation, coastal erosion and coastal inundation can affect the course of any river system.

## 3.2 FLOODING

Flooding (Figure3.1) can be recurrent and typical in the lower delta regions. Flooding can form in marine and riverine ecosystems. Both these ecosystems are vulnerable to geohazards. It is essential to make an integrated study of the river basins. We need to collect baseline data from the remotely sensed data, historical and archival data and field surveys.

**Figure 3.1** Flooding.

**Source:** https://www.theguardian.com/us-news/2019/jun/03/so-much-land-under-so-much-water-extreme-flooding-is-drowning-parts-of-the-midwest

Through remotely sensed data we can know land use extent, lineaments, structure, channel courses, shoreline and geomorphology. We need to go for field checks and maps (Edward., 2018). Through historical and archival data, we can get the timelines of the major infrastructures, agriculture, settlements and industries. This data can help us to estimate the nutrients and effluents. We can use field surveys to obtain paleochannels, shorelines and micro sedimentary properties. The natural water in the earth can be important in regulating biogeochemical cycles (Doddy et al., 2009). The floodplains can be considered as integrated components of any sort of river system where alluvial type is dominant. In general, water will be flowing in a designated channel and mostly the flow will be in the boundaries and it will be outside the limits only if the natural processes force the water content. Sometimes, water can flow upon the banks and this will result in the inundation of the land near the rivers. These can be called floodplains. This can be classified as a natural hazard (Robert et al., 2007). Due to this, we can expect severe damage to human settlements and infrastructures. In some cases, loss of life can be recorded too. This will be based on the magnitude of the forces incurred by the natural processes. The floodplains can be any part of the river system that can be observed in case of rare hydrological events. This can occur if the water flow exceeds channel limits. Technically, the floodplains can diminish the speed of water flow whilst flood events. They can help in diverting the water flow into a wider expanse. The flow that is observed in the floodplain can be shallow. The depths are too low to ensure that the high velocities are checked from time to time. Most of the floodplains can be vegetated with shrubs, grass, and trees. The vegetation can reduce the flow velocities. The flood plains can attenuate the peak discharges that might occur in certain watersheds. The floodplains (Figure 3.2) can also exhibit some essential functions (James et al., 2015).

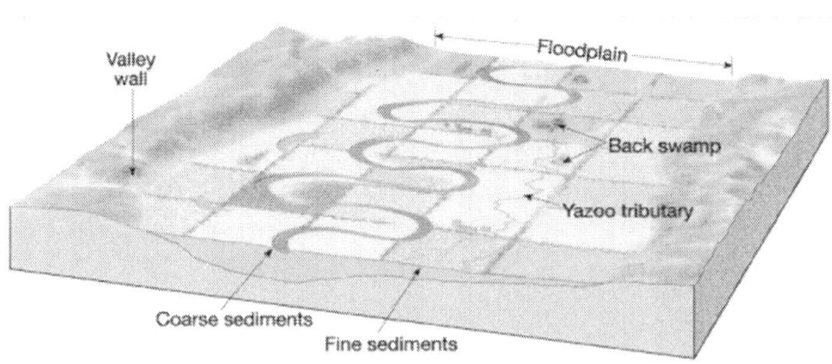

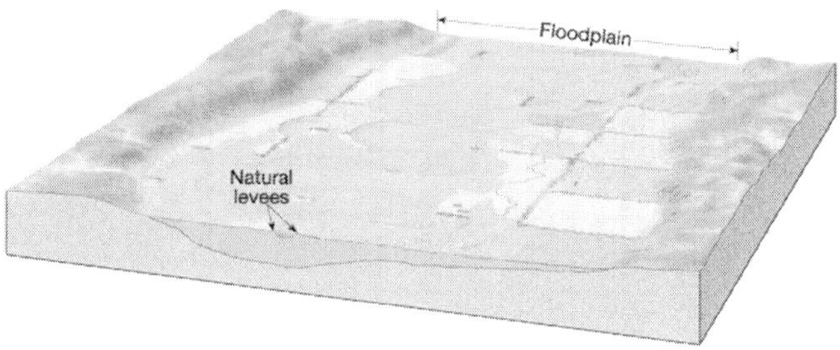


**Figure 3.2** Floodplains.

**Source:** https://www.theguardian.com/us-news/2019/jun/03/so-much-land-under-so-much-water-extreme-flooding-is-drowning-parts-of-the-midwest

The management of natural resources and water in a specific river basin can become a tedious task and it is based on the population explosion. Watershed managers can slow down the degradation of natural resources. They will strive hard to keep up the sustainable environments. The basic needs of the population at large can be addressed with the effective management of natural resources. Integrated river basin management and integrated watershed management (Figure 3.3) can be combined to form integrated natural resources management.

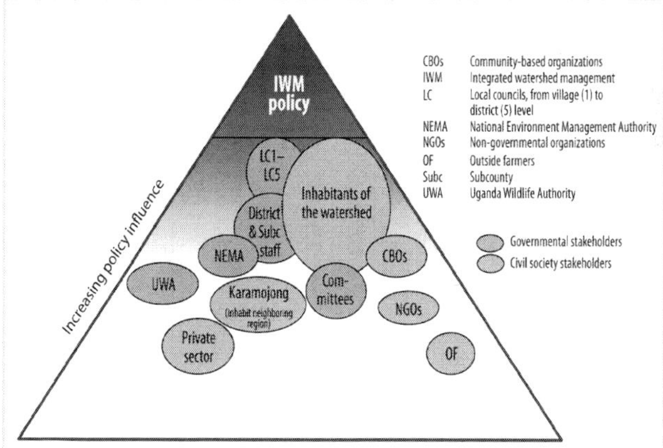


**Figure 3.3** Integrated watershed management.

**Source:** https://bioone.org/ContentImages/Journals/mred/33/2/MRD-JOURNAL-D-12-00031.1/graphic/i0276-4741-33-2-122-f05.jpg

The development, management of sustainable production systems can help us to maintain sustainable ecosystems. Watershed can be assumed as a catchment by several researchers (Janusz et al., 1985). It is nothing but a representation of the total terrestrial area that is being fed due to the drainage of water from a specific point. We can assume a watershed as a basic hydrological unit. It can be assumed as a social, political and economic unit in the case of management and planning of the natural resources (Thanh., 2017). We can also consider it as a physical cum biological unit. A river basin is nothing but a watershed but the scale is larger. The river basins can comprise all the watersheds that can feed them with water and other sediments. Integrated watershed management can deal with the interrelationships between water resources, soil and land use activities. We can also consider downstream areas and upland watersheds. The relationships between water and soils with natural resources can alter the course of the watersheds in the local areas. Mostly, the flow of water will be downhill and this doesn't need any political or social barriers. Integrated watershed management can help us in framing an effective strategy to practice sustainable natural resource management.

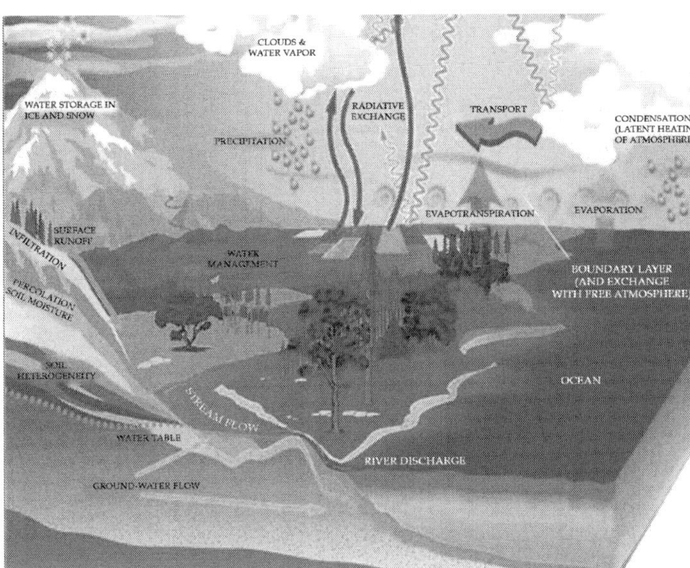


**Figure 3.4** Management of watersheds.

**Source:** http://lifegreensystems.com/wp-content/uploads/2014/05/Picture1.png

The management of watersheds (Figure 3.4) may not completely depend on biological and physical characteristics. The precipitation can get infiltrated into the soil horizons. It may not contribute to soil erosion in excess. If the watershed is in bad condition, we can expect degraded conditions. The integrated watershed management practices can ensure sustainable flows of water with high quality. We need to plan and implement economically and ecologically sound management strategies. The diminished water flows, stormflow peaks and streamflow volumes can be affected by unintentional and natural hydrological problems, road construction activities, wildfires, vegetative conversions, logging activities and grazing practices. The internal or inland drainage systems can be called endorheic systems. They are classified as closed hydrological systems. Most of the surface waters can drain into the terminal locations that are inland. In this case, water can seep into the ground and it can evaporate. It is important to manage the transboundary basins to restrict the complexities incurred in managing the limits of the river basins. The river basin can be any area classified as terrestrial and the surface run-off will be via lakes, rivers and streams. The river basin can be classified into elementary basins and they can be called sub-river basins (Ludwig., 2018).

The sustainability range of the development projects can be considered as an academic premise rather than an application interface. This aspect of simplifying the concept to minimize publicity about the due problems of natural ecosystems should be stopped. The management processes must take care of all the natural resource protection efforts as and when needed. There is a need for the systematization of the watershed management process. First, we need to collect information about watersheds and their characteristics. We need to collect the reports, photographs, maps of watersheds. The main use of this process is to understand the local conditions completely. It is essential to understand the routes of communication with the public residing in the area under study. We need to know the factors that can affect our observation of the natural resources at the local scale. We need to identify the possible stakeholders that are linked with both the management and development of the watersheds in several areas. It is important to use the latest information technology platforms to update and share the details. We should not ignore the connections between the strategies, solutions, obstacles, objectives and problems that are associated with the stakeholders. If we do that, we may not solve the problems of the core areas. We need to enumerate the management actions along with the technical and direct actions. We need to know the direct and technical actions that can be forwarded using the maps prepared by respective communities and governing authorities. These maps can give us an idea about the total watershed characteristics (Ghosh., 2017).

Most of the studies about river systems will be based on the sub-watersheds (Figure 3.5) or micro watersheds (Figure 3.6). These studies may need data about the wastelands, natural forests, irrigated area, groundwater extent and drainage density. We can digitize the drainage patterns with the aid of geographic information systems. We can use Horton's law to assign the stream orders. According to this law, we need to consider basin length, number of streams, perimeter, area and length of the stream. We can also analyze the morphometric parameters of watersheds that are delineated using Horton's law. Some of the morphometric parameters that are to be considered are compactness constant, circularity ratio, elongation ratio, relief ratio, drainage frequency and density, bifurcation ratio, length of the stream, order of stream, etc.

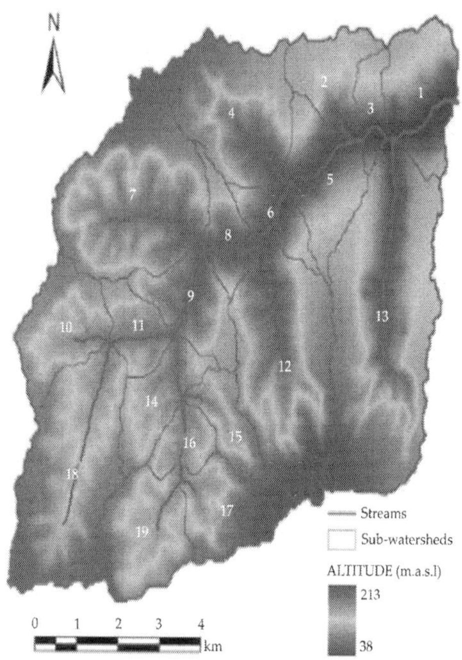


**Figure 3.5** Sub-watersheds.

**Source:** *https://www.researchgate.net/publication/305983450_Application_of_the_swat_model_in_the_Mamuaba_watershed_with_scarce_runoff_data/figures?lo=1*

**Figure 3.6** Microwatersheds.

**Source:** https://www.researchgate.net/publication/270759548_Micro_watershed_characterization_and_prioritization_using_Geomatics_technology_for_natural_resources_management/figures?lo=1

The stream order (Figure 3.7) can be measured with the help of hierarchical ranking. IPCC in 2007 has mentioned that anthropogenic activities can impact water resources at both regional and global scales. The urban centers can be translated as the sources of change in the river systems. The urban populace will have a limited water supply and this can be attributed to the population densities.

The meteorological and urban forcing can be witnessed with the changing water cycles. These water cycles will vary in the river basin scales. It is to be noted that the existing climate models may not be accurate in estimating the changing water cycles. These models will come with coarse resolution and if we need high spatial variability, we need to devise advanced methodologies. The water cycles at the river basin will be dynamic and to understand these cycles we need to use sophisticated sensors and other instrumentations. Most of the regional models can reflect the orographic features that may affect the water quantity and quality of the river basin. These models can also represent the land use dynamics. The atmospheric phenomena at local and regional levels like mesoscale flows, storms and sea breeze can be represented in the regional models at appropriate scales.

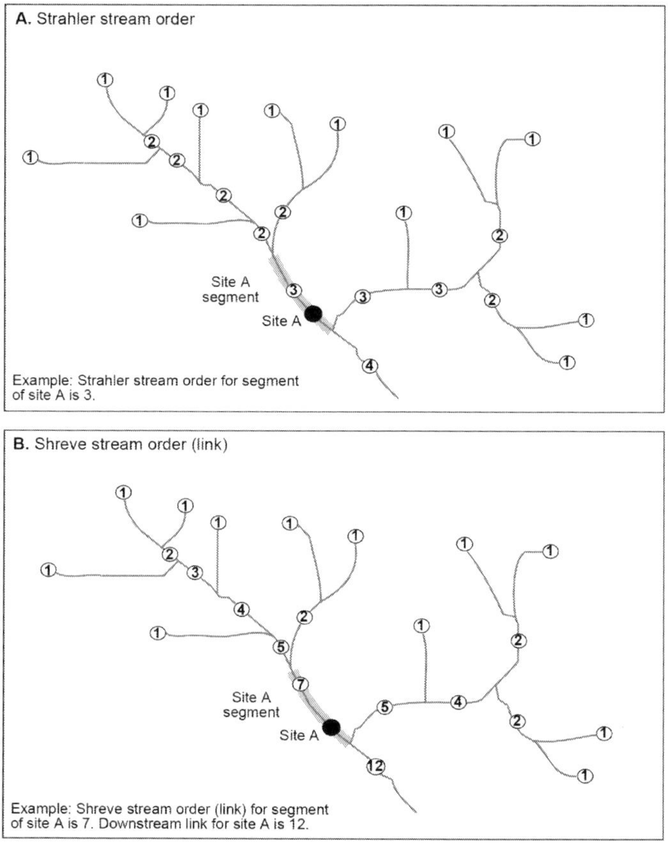


**Figure 3.7** Stream order.

**Source:** https://www.researchgate.net/publication/270589226_Revised_Methods_for_Characterizing_Stream_Habitat_in_the_NATIONAL_Water_Quality_Assessment_Program/figures

Freeze and Harlan proposed hydrological models (Figure 3.8) that can be used for basin scale. These models were focused on representing the vegetation changes, anthropogenic activities and land use dynamics etc. These models need inputs in the form of data aboutgeography, vegetation and soils. There are some models that consider water quality issues, agricultural activities and water supply conditions. Out of all the models studied, the physically-based hydrological models can be classified as effective in quantifying the land-use dynamics. We can use global climatic models or GCM to execute hydrological models.

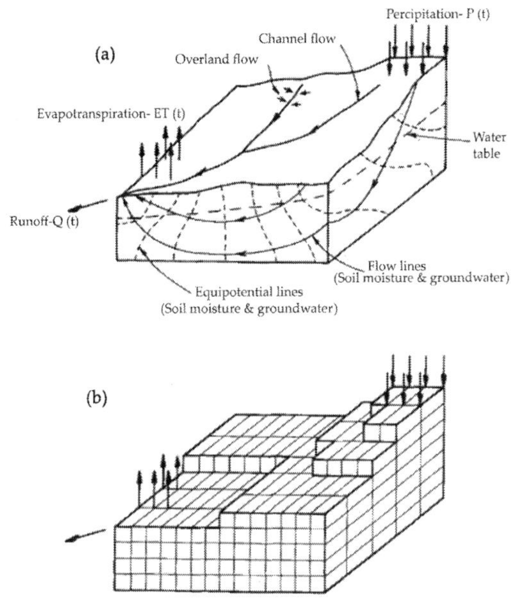


**Figure 3.8** Freeze and Harlan proposed hydrological models.

**Source:** *https://www.sciencedirect.com/science/article/abs/pii/S0022169419310443*

## 3.3 GCMS

These GCMs (Figure 3.9) can offer us at least 20 km resolution that can represent atmospheric forcing in the local regions (James., 2000). With these models, we can easily simulate the climate change effects on snowcaps followed by the discharge of excess water into the rivers. The distributed models can be used to implement water management plans easily. We can apply one-way coupling hydrological and meteorological models to obtain hydrological models. We can use precipitation and air temperatures in the form of hydrometeorological variables to simulate the regional climatic situations. The atmospheric simulation systems and numerical weather prediction systems can be used in weather research and forecasting with appropriate resolutions and accuracy (Akio., 2000).

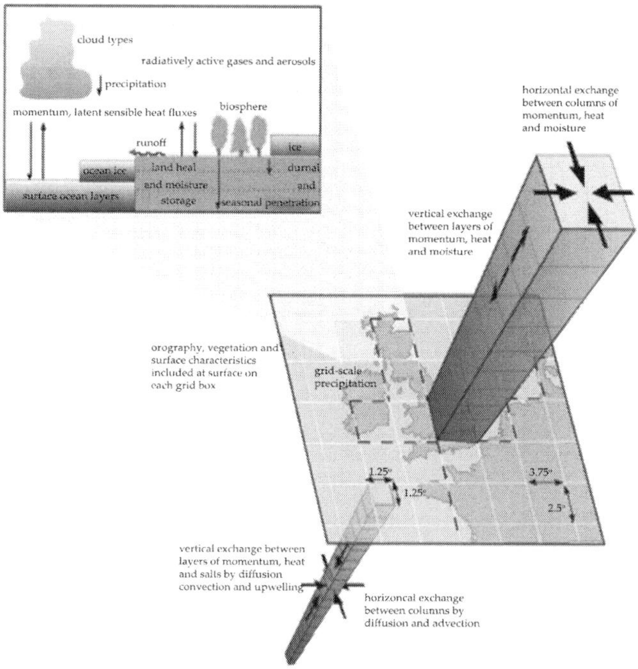


**Figure 3.9** GCM.

**Source:** https://www.ipcc-data.org/guidelines/pages/gcm_guide.html

The weather research and forecasting (Figure 3.10) or WRF can be used at global scales and local scales. We can use variables related to boundary layer, land surface, radiation and microphysics thereof. We can opt for distributed hydrological modeling at high resolutions. These modeling methods can be used in the simulation of river discharge and water flow to several hydrometeorological variables in place. The reservoir operations, water intake, groundwater flow, surface runoff and river flow can be modeled with the aid of distributed hydrological models (Paul., 2000). This will be based on the model devised to represent rainfall-runoff. This model can be called a hydrological river basin environment assessment model. We can use the Thiessen polygon model if there is an observation network in place to investigate the river flow levels. This method can be useful in simulating the orographic precipitation and terrain conditions. To execute an evapotranspiration model, we need to consider shortwave radiation flux, the pressure of water vapor, surface pressure, wind speed on the surface and air temperature. We can modulate these parameters in the form of grids. The modeled data can be used as inputs for any hydrological modeling process. We can also use the downscaled general circulation models as input (Phillips et al., 2021).

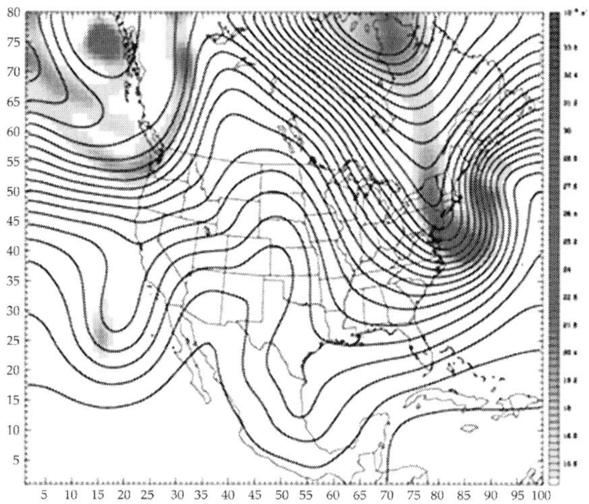

**Figure 3.10** Weather research and forecasting.

**Source:** https://narac.llnl.gov/tools/operational-modeling/weather-research-forecast-model

The environmental flows can be classified as critical tools in understanding the environmental distress of the river systems. This can be attributed to the scarce water resources. Managers need to understand the environmental objectives that affect the social and economic domains. The freshwater ecosystems are being degraded beyond threshold limits. There will be strong connectivity of hydrological resources in the river basins. These resources can affect the communities that can survive downstream. This can be achieved by sustainable water resource management strategies. The socio-hydrological management and strategic adaptive management can be considered as important aspects of river basin management. There should be the availability of data associated with the water resources so that the water planners can easily cope up with the changes that might be necessary. Firstly, we can start with the risk assessment or characterization. With the risk assessment, we can ensure and quantify the impact of climate change on the risks and anthropogenic pressures on the river ecosystem. We need to monitor and assess the status of the river ecosystem dynamics. We need to investigate the contemporary impacts of climate change via monitoring processes. Then, we can go for an objective setting. After this stage, we can go for economic analysis. It is essential to install certain measures and remediation programs to minimize the impact of climate change. Some of the physical factors that can affect climate change are effective water retention time, altered water discharge, elevated erosion and increased droughts. Some of the climate-induced changes that might occur in the river basins are increased toxic substances, groundwater salinity, reduced dissolved oxygen, discoloration of the surface waters and high-water turbidity levels. There will be limited growth of zooplankton, salmonids and macrophytes. There will be the growth of algal blooms that are very harmful.

Chapter 4

# Introduction to Reservoir Engineering

## CONTENTS


4.1 Introduction ........ 50

4.2 Reservoir Engineering ........ 50

4.3 Design ........ 53

4.4 Mineral Deposits ........ 54

4.5 Reservoir Rocks ........ 56

4.6 Logging Tools ........ 56

4.7 Permeability ........ 57

4.8 Well Logging ........ 60

## 4.1 INTRODUCTION

Reservoir engineering (Figure 4.1) can be classified as a part of petroleum engineering in general. It mostly comprises the effective management of both gas and oil reservoirs (Hugh., 1988). Mostly, this subject will deal with the economic and technical sense of the reservoirs. The reservoir engineering subject can be categorized as a dynamic area of engineering. This subject area can be classified as an integrated discipline. It constitutes computer-based simulations, production, drilling, petrophysics, geochemistry, geophysics and geology subjects. It also deals with the environmental, economic and regulatory aspects. At the beginning of the 20th century, most of the petroleum production relied on onshore fields and it was easy to manage these fields. The reservoirs can be constructed based on conventional and unconventional technologies. The construction of reservoirs was based on the traditional techniques that considered fluid and rock characteristics. There are certain advances in the construction of reservoirs that demand innovative processes and economically viable conditions. These techniques can be used in reservoir engineering to dig to greater depths (Grant., 1988).

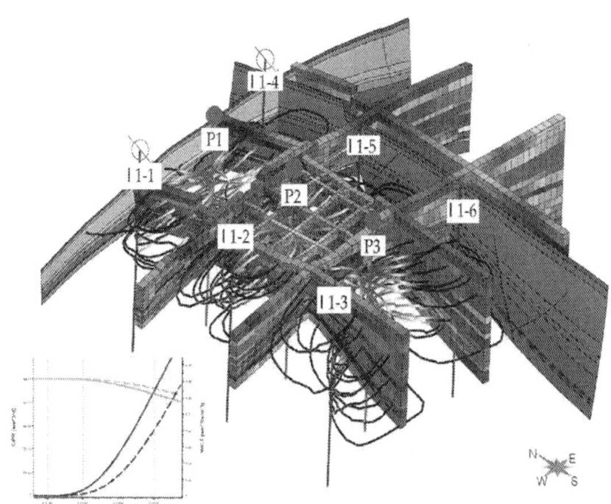


**Figure 4.1** Reservoir engineering.

**Source:** *https://idpe.org/reservoirs-engineering-services/*

## 4.2 RESERVOIR ENGINEERING

In reservoir engineering, we need to consider several rock characteristics. Some of them are the lithology characteristics that consist of sandstone, carbonate, coalbed, etc. The native configuration of the rocks is important. We need to consider source rock, reservoir rock, etc. The rocks might be tight, consolidated and unconsolidated. We need to consider geological complexity also. The rocks might be single-layered or multi-layered. We can observe stratification in the case of multi-layered rocks. They can be noncommunicating, partial communicating and communicating. In some cases, we can even find fractured rocks. There will be faulted rocks also. They can be non-sealing, partially sealing and sealing. We can also expect a compartmental type of rock formation. In some cases, there will be an immediate need to know the conductivity characteristics too. Most of the rocks can be heterogeneous (Porges., 2006). This can present diverse rock characteristics in this case. Location parameters can be important while planning a reservoir system. It can be onshore, offshore, shallow and deep. The pressure of reservoirs can be considered. It can be over pressured and under pressured and that depends on the inflow of the water. Reservoir drive energy can be considered in this case. This energy can be based on the thermal, external fluid injection, rock compaction, aquifer type, gravity, gravity, expansion rate of rocks and fluids, gas cap and depletion.

**Figure 4.*2*** Reservoirboundary.

**Source:** https://www.usgs.gov/media/images/boundary-reservoir-0

The reservoir boundary (Figure 4.2) can be classified into bottom-water drive boundary, edge water drive boundary and closed boundaries. We need to estimate reservoir dip also. There will be a steep inclination in certain cases. This will help us in locating wells. The production modes can be considered. The mode of production can be classified into primary, secondary and tertiary classes. The primary mode mainly constitutes the production of natural reservoir energy. The secondary mode of production involved the augmentation of water flooding. The tertiary mode of production involves the injection of chemicals, thermal treatment and foam. Some of the production characteristics that can be considered are single-phase and multiphase flow. The single-phase flow may involve gas or oil phases. In the case of multiphase flow, we can consider water/gas, gas/water, oil/water and oil/gas. We can expect a high gas to oil ratio and high water cut in this case. The life of the reservoirs can be important. It starts by describing the initial production stage. This step is followed by peak production, lowered production and a reservoir that is matured. It is to be known that no two reservoirs share similar features. Every reservoir can exhibit peculiar features. This can involve integration of reservoir data, interpretation, validation, computer-based fluid flow models, visualization of the processes involving fluid flow and characterization of the complexities involved in the geological formations (Tarek., 2010).

It is essential to always have an understanding of the entire mechanisms of reservoirs. It constitutes both conceptualization and visualization of the fluid flow and rock characteristics. The reservoirs that were tagged as unconventional reservoirs can present certain new challenges. We can integrate data produced while constructing reservoirs with petrophysical, geological and geophysical parameters. This data can be used to create a conceptual and acceptable reservoir model. We can estimate gas and oil using different methodologies. This will also include flow characteristics. We need to simulate and develop computer-based models to estimate the performance of the reservoirs. It is essential to go for reservoir surveillance. This will help us produce charts and reservoirs. This knowledge can help us design and predict future courses of action.

It is essential to work closely with the scientists and engineers to get a complete idea about the reservoirs. We need to adhere to the best management practices to get a complete understanding of reservoir construction and management. We need to collect information about the reservoirs and the fluid flow. This will be based on the petrophysical, geophysical and geological studies done previously. We need to estimate the gas and oil reserved depending on the data available. We need to develop a plan to drill the new wells and this will be based on several factors. We need to drill and design the wells according to expert advice. It is essential to estimate the water flood dynamics to the local geological setup (Iqbal et al., 2016).

## 4.3 DESIGN

We need to design and carefully implement the recovery process. In some cases, there will be the production of oil along with the release of water. We need to detect zones that can produce water. The fields can be rejuvenated as per our needs. We need to opt for certain recovery operations that can easily be done. This process must be economically feasible (Mohammad., 2015). We need to design well tests according to the project plans and should be able to analyze data accordingly. It is important to collect data using petrophysical, geological and 3D seismic studies. These studies should be done regularly and without errors. It is important to characterize the performance of reservoirs. We need to monitor wells and also should be able to evaluate the recovery options. We need to combat the water inflow through the crevices of the geological formations. We need to select candidates that can easily come up with the results. It is always essential to update reservoir models. This must be done by taking into consideration the production process. We need to perform simulation studies as and when needed. It is important to regulate and monitor the reservoir performance periodically. We need to always upgrade the reserve estimates in line with the latest developments. The walkthrough process can be used too. In this process, we need to start with the correlations, regional trends, existing wells, newly drilled wells, geomechanical, geochemical, geological and petrophysical data. It is always important to design and investigate the wells using diverse tests. These tests can be used to define the properties of the reservoirs. The characteristics of the hydraulic fracture can be represented accordingly. We need to design and automate the optimization of the hydraulic fracturing process. This will guide us to the completion of the wells (Philip et al., 2014).

We need to build and validate the reservoir models. It is essential to simulate the performance of the reservoirs. This will help us generate diverse scenarios. All the scenarios will be dealing with the hydraulic fracturing and stages involved in the entire process. We need to do economic analysis and forecasts must be done with the aid of advanced programs available. We need to design the worst-case scenarios with utmost confidence so that any adverse situations that are linked with the life cycle of the reservoirs can be mitigated. After all the necessary prerequisites are met, we need to produce a field development plan. This plan must be subjected to the prior approval of the stakeholders. Most of the theoretical and practical considerations on reservoir engineering will deal with the fluid properties and reservoir rocks. Mostly, the reservoirs that are to be installed will be based on the behavior of the fluid phase. The reservoir engineering will be useful to manage the wells. This can be used in the spacing of the wells and locating them according to the needs.

## 4.4 MINERAL DEPOSITS

Shale can be classified as an important rock that can be available in large amounts. It is mainly found in sedimentary basins. Most of the rock volumes are made up of shale. The conventional reservoirs will be made up of carbonate and sandstone formations. All these elements will be integrated within the shale deposits. Mostly, the carbonate reservoirs will be formed. We can find petroleum reservoirs in the carbonate reservoirs. The fluid and rock properties can provide us an insight into the reservoir characteristics that can be essential in the analysis and development of reservoirs and predict their life cycle. The rock properties can be classified into static properties and dynamic properties. The static properties can be subjected to environmental changes and this can be called a depositional environment. These static properties can withstand different geological events. This environment can affect the rock characteristics and rock types (Figure 4.3).

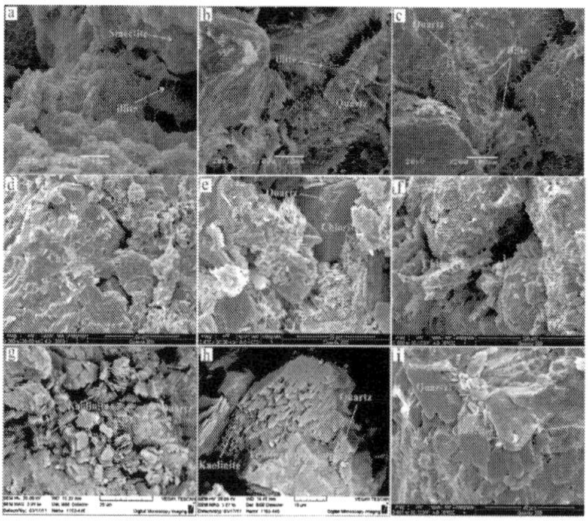

**Figure 4.3** Mineral deposits.

**Source:** https://www.mdpi.com/2075-163X/9/7/436/htm#

Some of the static properties of rocks are porosity, size and distribution of the pores, diameter of the pore throat, permeability and compressibility of the rocks. It is important to consider two important properties of rock, they are porosity and permeability. These properties are important in storing and generating the water flow. The void spaces will be microscopic and some of the water will be accumulated in these spaces. The tendency of the rocks to allow fluids to flow in the form of conduits and pathways can be called as permeability of that rock. The dynamic properties of the rock can be affected by both fluid and rock properties that may exist in the reservoirs. Some of the dynamic properties are wettability, capillary pressure, saturation of fluids and relative permeability. These properties will change with time as the inflow and outflow of water will vary. There are some geochemical properties that need to be considered. They are total organic carbon and vitrinite reflectance. It is important to understand the geomechanics of the rock structures. This gives us an idea about hydraulic fracturing. We can understand the firm formation of artificial fractures and natural fracture networks. We need to consider fracture stress, Poisson's ratio and Young's modulus. The reservoir rocks can present a network of pores that are microscopic. At these pores, the fluids of the reservoir will be stored. The volume of void spaces and pore volume/bulk volume of the rock that is porous can be called porosity of the rocks under investigation. The volume of the rock matrix that is solid and pore volume can be described as the bulk volume of the rock. This can be called the grain volume of the rock also. It is to be noted that all the pores may not be interlinked in the rocks that are present in the reservoir. It is essential to know the effective porosity of the rocks. The percentage of absolute porosity can be defined as the ratio between the volume of voids and pores in the rock to the bulk volume of the rock multiplied by 100. Generally, the effective porosity can be less than the absolute porosity. This is due to the fact that the effective porosity can be attributed to the pores that are interconnected. The knowledge of the porous network that is interconnected can be important. In the case of dead-end pores, there will be limited porosity and hence may not be considered in determining the porosity of the rocks.

The porosity present in the rocks can be categorized into primary and secondary porosities. The porosity that might have developed initially in the rock when it is in its native state can be classified as primary porosity. The secondary porosity can be developed due to the geochemical and geological process as time progresses. There will be deposition mainly due to secondary porosity. The deposition that might occur due to secondary porosity can be explained by various factors. There will be dolomitization and leaching of the carbonate rocks. This can be due to the action of chemical solutions. Due to these reactions, there will be the formation of cavities and vugs. We can also observe fractures and fissures that are microscopic. Mostly, the secondary porosity can be present in the carbonate rocks. In some cases, we can observe secondary porosity in the sandstones and this can be due to other external factors in place. Most of the reservoirs that yield petroleum can be formed as dolomite formations. Most of the heterogeneities that might occur in the reservoirs can be attributed to the secondary porosity. This can always produce uncertainty in the analysis of the reservoirs. We need to consider sorting and grain size to determine the porosity.

## 4.5 RESERVOIR ROCKS

The reservoir rocks that are tagged as unconventional can have shale deposits. There will be low porosity in the case of shale-dominated reservoirs. Some of the geological formations can present joints, fissures and fractures that resulted due to diverse stresses. The porosity levels can vary from one reservoir to another reservoir. We can get the porosity levels of wells using petrophysical studies and this will be based on the core and log analysis.

The average porosity levels can be used in the reservoir analysis. In this case, the rocks that are present in the reservoir should be homogeneous. We can use several mathematical algorithms to calculate porosity at various locations where data is not available. We can use nuclear magnetic resonance and drill for studying the wells and reservoirs in detail. We can measure the effective porosity by using the core sample in dry conditions in the fluid. The fluid density should be known prior to the analysis and measurements. The grain volume and bulk volume obtained from the core sample can be estimated using the displacement method. In this process, we can rely on the core sample that is previously immersed in water. We need to estimate the amount of water that is being displaced. We can investigate porosity with the help of a porosimeter. In this process, we can place the core sample that is in dried condition at the vacuum chamber (Gian., 1994).

## 4.6 LOGGING TOOLS

The logging tools can be used in defining the porosity. Some of the logging tools are based on neutron porosity, density porosity and acoustic properties. The tools that were installed in the boreholes can be used to investigate the rock properties. To get these properties to scale, we need sources and receptors. The receptors can be more than one. We can use sonic logs and acoustic logs to investigate the properties of the rocks that can be used in reservoir construction. The sound waves that will be used can travel slowly in the porous rocks that were saturated with the fluids. Most of the density tools will use gamma rays. These rays or radiation can be used to probe the electrons present in the fluids and rocks. This probing can help us in determining the pores present in the rocks. Using these techniques, we can define the electron density of the subsurface regions. The electron density obtained through instruments can be proportional to the density of the rocks that are filled with the rocks (Alumbagh., 2002). The porosity values can be estimated with the aid of density values of water, dolomite, limestone and sandstone. In the case of neutron porosity tools, there will be the production of neutrons of high energy. These neutrons can be used to know the formation of subsurface structures. In presence of pore fluids, the neutrons applied will react with the hydrogen atoms. We can use nuclear magnetic resonance to investigate the hydrogen nuclei that are present in the pore fluids. These tools can be used to know the interaction between the H nuclei of the fluid with the surface of the rock grains. This might not be possible with the larger pores. The LWD approach or Logging while drilling operation was classified as an old technique as far as reservoir engineering is considered. There are many sensors and tools that can be used in the assembly of boreholes whilst drilling. We can use neutron porosity and sonics to measure porosity. LWD can supply real-time data.

## 4.7 PERMEABILITY

The permeability (Figure 4.4) of the rocks can give us an idea about the fluids present in the reservoirs that can migrate from the pores. The high rock permeability levels can be useful in enhancing recovery efficiency. The productivity levels will be increased proportionally.

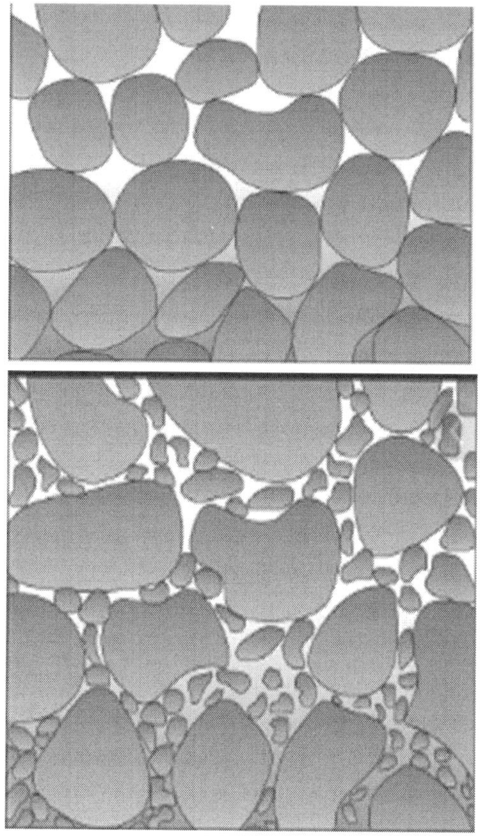

**Figure 4.4** Rock permeability.

**Source:** *https://www.pinterest.com/pin/781093129095836282/*

The permeability levels can be calculated using Darcy's law. The fluid flow rate can be proportional to the hydraulic head of water. There will be a proportionality constant that can define hydraulic conductivity. This hydraulic conductivity can be observed in the case of a porous medium. We can measure low permeability using a pressure decay method in the case of shale reservoirs. The vertical permeability will be lesser than the horizontal permeability. The grains can be observed in the form of sediments and flaky particles. Mostly they will be laminated. There will be several depositional sequences as time progresses. We can observe both vertical and horizontal permeability. Both vertical and horizontal permeability can be considered as an important factor in reservoir engineering. There will be high levels of vertical permeability and this will lead to a breakthrough of water. We can observe water slumping in case of flood events and this can be due to the high rate of vertical permeability. The permeability of rocks will be based on certain directions of the reservoirs. The heterogeneity of the reservoirs can be based on the streaks that might have been present in the reservoirs. We can observe certain intervals in the geological formations. These intervals can be conducive and the transport will be in lateral modes. The steaks that can exhibit high permeability rates can be challenging in the reservoir design. There should be sorting followed by the orientation of the grains (Habana., 2002).

The impurities, cementation rates of grains, pore channels and pore throat sizes can be considered vital factors that can affect the permeability rates of the rock structures. There are several factors that can be used to estimate the porosity of the rocks. The permeability rates may vary with the reservoir type. This can be attributed to the porosity of the rivers. The heterogeneities, cavities and microfractures can be considered essential factors that can define the permeability of the rock structures. All these elements can affect the performance of the reservoirs. The pore volume variation and the pressure in the reservoir can be used in knowing the formation compressibility of the reservoir. The rocks that constitute the construction of reservoirs can be compressible. There will be a limited decrease in the pore volume. The rocks that are made up of unconsolidated sandstone can be classified as highly compressible. The compressibility in the formation can be considered an important factor in reservoir analysis. This can be due to several factors. In the initial production stages of the reservoir, the reduction of pore spaces in association with the expansion of the fluids can add to the driving energy. There will be variation in the pore volumes if the pressure of the reservoir declines. This can be attributed to the compressibility levels. If the pressure of the reservoir decreases, then the fluid flow rate will be decreased in the compressible rocks. This can be attributed to the decreased permeability and rock porosity.

The grain size can be classified as an important factor that is to be considered. It is worth noting that the grain size alone may not affect porosity. The grains present in the rock should be uniformly arranged. The small size of the grains can result in the formation of pore channels that are narrow. This can result in reduced permeability. Most of the conglomerates can be identified by coarse grains. The grain size will be less in sand, silt and shale. It is important to know that the smaller grain sizes can reduce the size of the pore throat openings and pore spaces. The sorting of grains can be considered an important process. The rocks that have minimally sorted grains may present large pores. In some cases, these large pores will be occupied by smaller grains. Most of the pore channels can be small and this can result in low permeability. The sorted sandstones and clean sandstones can present permeability and porosity. Cementation of the grains can be important too. Due to small porosity, we can expect a high degree of cementation. The rocks that are cemented can exhibit less permeability. We need to consider the diameter of the pore throat. We can expect low porosity in the case of rocks that have minute pore throats. Due to reduced porosity, there will be limited permeability. The leaching and solution process can affect the porosity. The leaching of carbonates can result in secondary porosity. The permeability of the rocks will be increased due to leaching and solution. Mostly, the solution effects can be observed in the carbonates. We can observe this type of process in sandstones. The fractures can affect the porosity levels too. The fracture porosity can be important while describing the rock porosity as it is less than the matrix porosity. The reservoirs can exhibit dual permeability systems and dual porosity.

Interfacial tension and surface tension can be considered important factors in designing a reservoir. The surface tension and interfacial tension that might occur in the rock pores can alter the flow characteristics. The fluid saturation that can be present in the core sample can be estimated by evaporating and heating the fluids. These fluids will get condensed in the tube from which we can quantify the required parameters using the Dean-Stark method. The capillary pressure that exists in the porous medium can be affected by the saturation levels. There are many methods that can be used to estimate capillary pressures. Some of the methods are the Leverett method, mercury injection method, porous diaphragm method and centrifuge method. The mercury injection can be categorized as a rapid testing method to estimate the capillary pressure of a sample. It can be used to investigate the capillary pressure of the sample that is irregular in shape. The distribution of the pore size can be analyzed with this method. It is to be noted that the sample may not be used for further analysis. Mostly, the effective permeability will be based on the saturation levels of the fluids.

## 4.8 WELL LOGGING

Well logging (Figure 4.5) can be used in several fields of mining. This technique will be used to quantify the rock properties with the aid of sensors arranged at various locations of bore holes. Some of the rock characteristics that can be considered are the degree of fluid invasion, fluid saturation, porosity, geological structure and lithology. Most of these sensors can use gamma radiation, neutrons, acoustics, electromagnetics and electricity. These sensors can send and receive signals that might be bounced back from the geological formations (Liu., 2016).

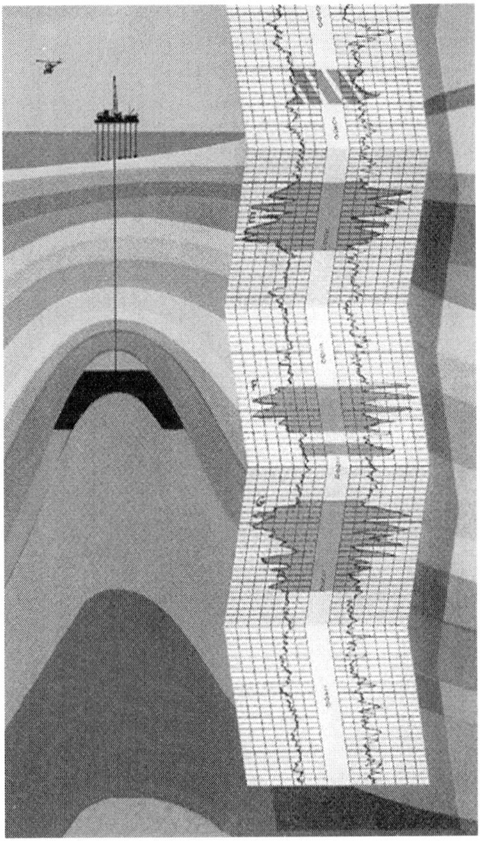

**Figure 4.5** Well logging.

**Source:** *http://www.sjvgeology.org/oil/well_log.jpg*

There are certain imaging and logging tools to characterize reservoirs as they can offer information about faults and fractures. The logging process can be classified into a cased hole and an open hole. With the collection and analysis of the open hole logs, we can get information about the hydrocarbon intervals. The integrity of the casing can be estimated.

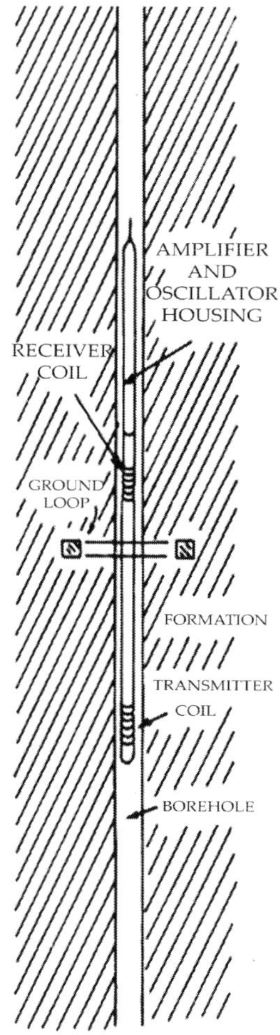


**Figure 4.6** Resistivity logs.

**Source:** *https://wiki.seg.org/wiki/Resistivity_log*

Resistivity logs (Figure 4.6) can be considered in the logging process. Water can be classified as the best conductor of electricity. These tools consist of electrodes. These electrodes can send signals to the formations that are filled with fluids. The current will reach the second electrode to complete the circuit. Based on the conductivity presented by the fluids that are in the geological formation, the intensity gets varied. We can use density logs to estimate the bulk density of the geological formation and this density logging tool uses radioactive sources.

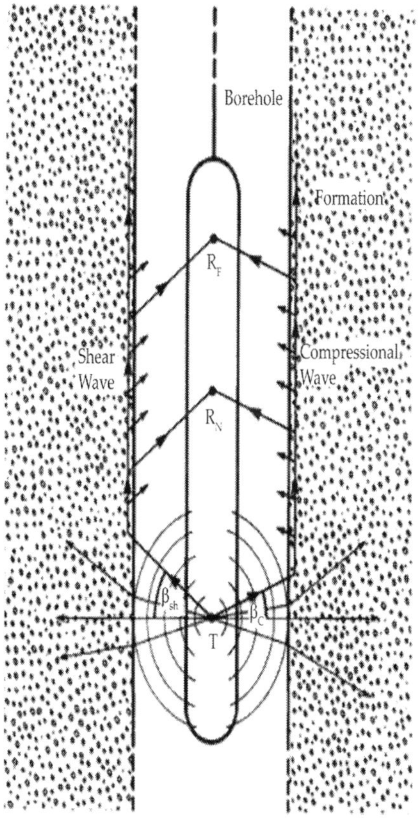


**Figure 4.7** Acousticlogs.

**Source:** https://www.sciencedirect.com/topics/earth-and-planetary-sciences/acoustic-logging

Acoustic logs (Figure 4.7) are also important in characterizing the reservoir rocks. These logs will be produced by sensing through sounding. The sounding process can be affected by texture, lithology and rock porosity. In this case, the transmitter will generate sound waves. We can produce these logs by estimating the time interval. These logs can be used to diagnose the rock features.

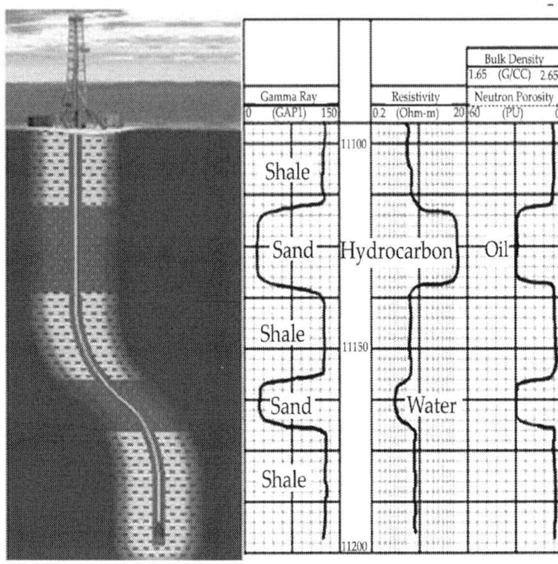


**Figure 4.8** Gamma ray logs.

**Source:** https://www.sciencedirect.com/topics/earth-and-planetary-sciences/acoustic-logging

Gamma ray logs (Figure 4.8) are another important set of tools that can be used in this process. This gamma radiation can be used to estimate the radioactivity of the formation and the observed radioactivity will be natural. We can observe radioactive potassium in the clay of the shale beds. These layers can be demarcated from the sandstone layers. These shale beds can be made up of quartz particles that can be non-radioactive. It is to be noted that shale beds will constitute thorium and uranium. Caliper logs (Figure 4.9) can be used to know the profile of the boreholes. We can obtain the shape and diameter of the boreholes using caliper logs. We can use spectral noise logging as an acoustic tool to investigate the injection intervals and prediction intervals. This tool uses noise that will be generated from the fluid flow. We can also investigate the leaks. The heterogeneities of the reservoir can be megascopic, macroscopic and microscopic. The microscopic heterogeneities will comprise pore throat diameter and pore size. There will be minute channels that are formed in the rocks. We can also know the impurities and grain sorting using microscopic heterogeneities. If the rock type is sandstone, we can observe grains that are poorly sorted. In this rock type, we can observe certain impurities generated from mudstone and siltstone. The sandstone rocks can also degrade reservoir quality due to the presence of rock heterogeneities. The fractures that are present in the sandstone can also increase productivity.

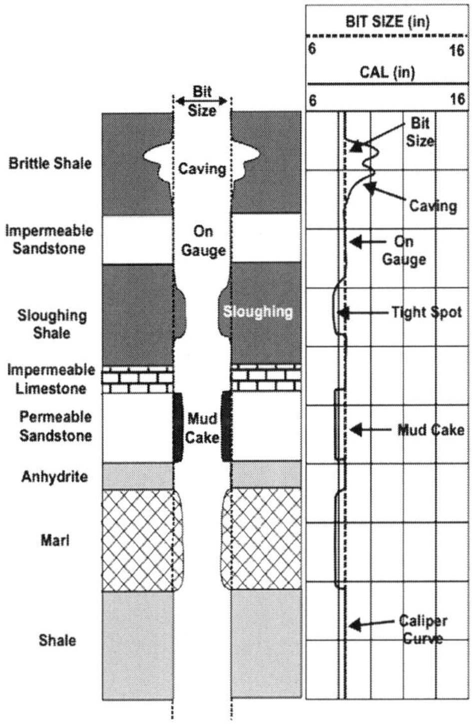


**Figure 4.9** Caliperlogs.

**Source:** https://oilfieldteam.com/en/a/learning/caliper-logs

The productivity will be higher in the tight sandstones. In the case of carbonate rocks, there will be the presence of fractures, solution cavities, channels and vugs. The channels and fractures that are formed will increase the primary production. Water breakthrough is common in these rock types. In the case of shale rocks, we can observe fractures and laminations. There will be important variations in the petrophysical, geomechanical and geochemical properties. In this case, we can expect good performance but it will be unpredictable. The thin section and core studies can bring about microscopic heterogeneities into scale. The rock properties can be upscaled in the case of reservoir studies. The micro seismic studies and tracer studies along with the pressure transient testing can be used to know the macroscopic structure of the rocks. This can be done with the well logging techniques.

Chapter

5

# Introduction to Water Resources Engineering

## CONTENTS

5.1 Introduction........68
5.2 Closed Conduits........71
5.3 Evapotranspiration........76
5.4 Rivers........80
5.5 Stream Orders........81
5.6 Forecasting........82
5.7 Artificial Intelligence........85
5.8 Models........86

## 5.1 INTRODUCTION

The water resources engineering (Figure 5.1) can be considered as an important branch of engineering that deals with the system design and regulation of water distribution. The timing, quality, and quantity of water supply to the human needs can be studied in this branch of engineering. The social, political, financial, economic and legal aspects of water resources will be discussed in this chapter. All these viewpoints can be considered important in the development process. Water that is present in water can be present in oceans and land. Hydrological cycles can be important to know the pathways of water both on land and atmosphere. The hydrological cycle can start with the water evaporation that occurs in the oceans. The evaporation of water will be based on the amount of sunlight that falls on the oceans. There will be convection and condensation of water in the atmosphere resulting in the formation of clouds. The vapor deposited in the atmosphere will precipitate on the ocean and land surfaces. This can be visible in the form of snow and rain. Rainfall that gets deposited on the land can be intercepted by the vegetation that grows on the land surface. Some of the water will be stored in the depressions formed on the land surface (Francisco., 2014).

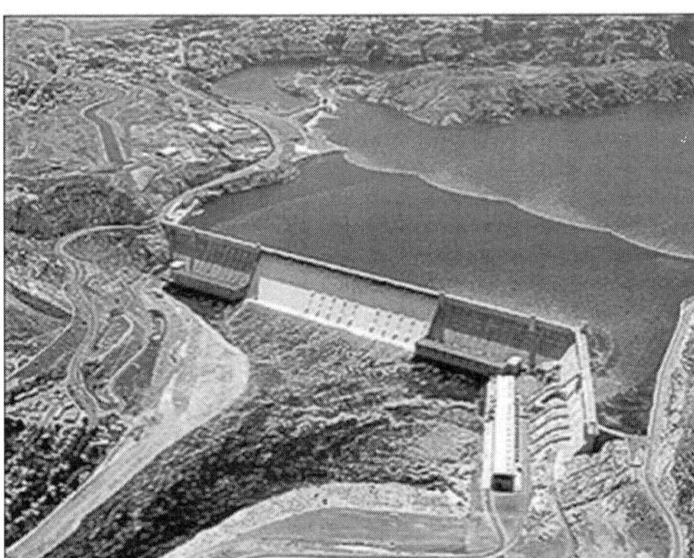

**Figure 5.1** Water Resource Engineering.

**Source:** *https://sites.google.com/site/civilengineeringforyou/jobs/water-resources-engineering*

There will be infiltration of water (Figure 5.2) in some areas. In some areas, water can enter the drainage channels and this can be due to surface runoff. There will be certain contaminant classes that are to be studied. The waste that is oxygen demanding can be produced from animal and plant wastes. We can get infectious agents like viruses and bacteria from water. Several plant nutrients can yield greater amounts of phosphates and nitrates.

**Figure 5.2** Infiltration of water.

**Source:** *https://www.usgs.gov/media/images/a-small-stream-georgia-disappearing-a-cave-infiltration*

The organic chemicals that are transported are grease, oil, detergents and pesticides. Some of the inorganic chemicals that can be transported via water are acids generated from the drainage of coal mines. Some of the inorganic chemicals from the steel plants comprise iron in larger amounts. The sediments that are formed due to land erosion can be transported via water. There will be a deposition of clay silt on the beds. This will lower the survival rate of the life forms. Most of the life forms may survive at the solid-liquid transition zones. In some cases, there will be the transport of radioactive substances. These radioactive substances can be produced from the mining industries. We can expect most of the radioactive products from the radioactive isotope processing industries. Heat can be generated in the industries and this will increase the heat of the surrounding water. Most of the cooling water used in the industries will be released into the local water bodies. These abnormal temperatures can damage the local biota. The natural variability and uncertainty will be present in the hydrological processes. These reservoir systems can be designed by considering the risk involved. We can design these systems based on critical event design, rick-based design and frequency-based design. The water resource systems can be classified into water use systems and water control systems (Jacques., 1996). The water control systems can be designed to regulate the spatial-temporal runoff of the rainfall deposited on land. It is essential to install flood control structures.

**Figure 5.3** Impoundments.

**Source:** *https://www.federatedenvironmental.com/wp-content/uploads/Surface-Impoundment.jpg*

The impoundments (Figure 5.3) of storage can drastically lower the flow rates of drainage channels, rivers and streams. The drainage system can control the water flowed from the industrial and domestic water supply. The flood control systems can be associated with the treatment of wastewater. The salinity control systems can be linked with the irrigation water-use systems. The sediment control systems can be associated with the production of hydropower. Floods can be described as the enhanced flow of water that can exceed the bearing capacity of the drainage channel or stream. Flood stages can be described as the elevation point where excess water tends to flow. Generally, this elevation point can be installed for drainage channels or streams.

The flood plains can be described as any dry land that is present nearby oceans, bays, lakes, streams and rivers. These flood plains will be inundated in case of flood situations. The floodplains are being encroached and this irregular practice can enhance the floodplain extent. The water regulation systems can include certain sewer systems that can be used to gather/collect and transport the runoff. We can also use storage reservoirs to reduce the peak runoff rates. This will also lower the pollutant loads, especially in the drainage channels. Most of the water-use systems can be used to help human habitations. Generally, water-use systems can be classified into irrigation systems, wastewater-treatment systems and collection systems and distribution systems. Water will be supplied to these systems from groundwater, lakes and rivers. The design capacity of these systems can be regulated by the population dynamics. It is also based on the industrial demands and characteristics of the area. Any designing aspect of water resource systems can be associated with different agencies. It is essential to collect geological and hydrological data.

## 5.2 CLOSED CONDUITS

The flowing fluid can fill the conduit and it is important to consider closed conduits. The closed conduits (Figure 5.4) can represent different cross sections. It will be based on the size and shape of the respective conduits. These conduits can be made up of different materials. We can use the Darcy-Weisbach equation to estimate the friction between the pipe boundary and moving fluid. The water distribution systems can aid in transporting water from the natural water resources to the treatment plants. The main components of these systems are meters, valves, storage facilities, pumps and pipelines. The water supply systems must ensure that the water demand of the public is met. Many methods need to be applied in forecasting the water demands. Some of the models that can be used are land-use models, multi-regression models, disaggregation models, extrapolation models and per-capita models. The Land-Use models can forecast the demands of public lands, industrial, commercial and residential areas. In this model, there will be separate projections for each segment of land-use. The multiple-regression models can be classified as empirical models (Lynn., 2013).

The per-capita forecast model can be used to know the population projections and water usage. I will consider the water demands on average. There are many sectors that can affect water demand in public, industrial, commercial and residential areas.

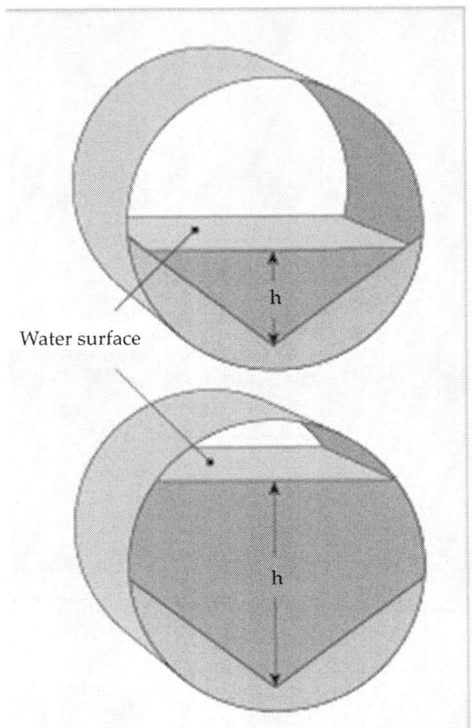


**Figure 5.4** Closed conduits.

**Source:** *http://stormwaterbook.safl.umn.edu/water-budget-measurement/flow-conduits*

The water distribution systems can comprise pipelines that can be used to transport the drinking water. The pipelines can comprise distribution mains, arterial mains and transmission lines. The water mains can be installed in the streets and ends of the water distribution systems. The improper installation of these mains can produce polluted water. The dimension of the water main can describe the carrying capacity. It is always important to select the sizes that can be used in case of high-water demand. The service lines can be pipes that comprise accessories to the main supply. The water surface will be exposed to the atmosphere in case of open-channel flows. We can find open channel flow in the case of rivers and canals. We can label the open channel flow and free surface flows. The open channel flows can be considered as complex flow types than the closed channel flows. Surface-water hydrology deals with the several properties of water from the rainfall, its movement and distribution. The surface runoff will be regulated by the land area at a specific location. It will be determined by the topography and shape of the land. This area can be called a watershed. We can call any area that falls in the watershed area and can absorb water from the rainfall event can be called a catchment area. The catchment area and drainage basin can be called a drainage area and drainage basin. The runoff that is generated from the catchment and watershed will be high in the catchment outlet. This can be called a pour point. Some of the characteristics of this area can determine the quality and quantity of the runoff. This can be important in the case of rainfall events. Most of the precipitation can occur in the form of snow and rain. We can expect sleet and hail in rare cases. Most of the drainage systems can regulate the runoff that is generated from rainfall events. Most of the precipitation can result due to the uplifting of the moisture filled air masses in the atmosphere. As the vapor reaches greater heights, these air masses will get cooled and hence the condensation (Figure 5.5) process will get initiated.

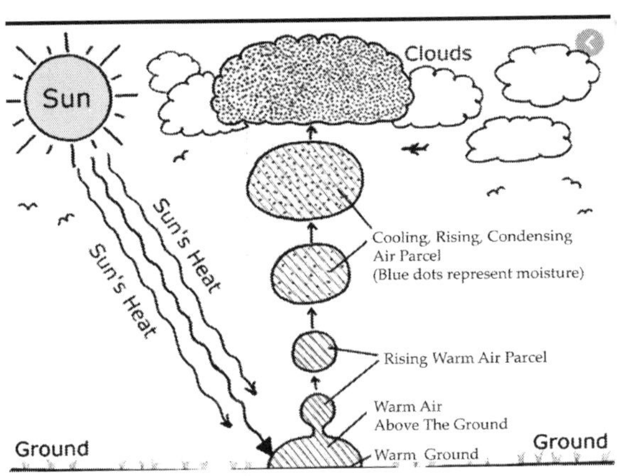


**Figure 5.5** Condensation of clouds.

**Source:** *https://climate.ncsu.edu/learn/*

There are certain conditions that bring about rainfall events. There will be the cooling of the air mass. Some of the water droplets can get condensed with the help of nuclei. The water droplets can grow in size. There will be certain mechanisms that can enhance the density of the droplets (Bear., 1982). Due to the increased density of the droplets, they fall onto the ground. The cloud droplets mostly get generated using nuclei meant for the condensation process. These nuclei will be less than 1 micrometer. Most of these nuclei will comprise combustion garbage, dust and sea salts. We can call these particles atmospheric aerosols. These aerosols can play a vital role in controlling the solar radiation that falls on the earth's surface. If the air is classified as pure, then there will be condensation of water vapor. This will result in the formation of water droplets. Air will be completely saturated with the water molecules. In some cases, there will be freezing temperatures and this will promote the formation of cold clouds. There will be the formation of ice crystals and water droplets. In this case, air will be supersaturated with the available water molecules. There will be the formation of ice crystals and the theory related to the formation of these ice crystals can be called the Bergeron-Findeisen process (Figure 5.6).

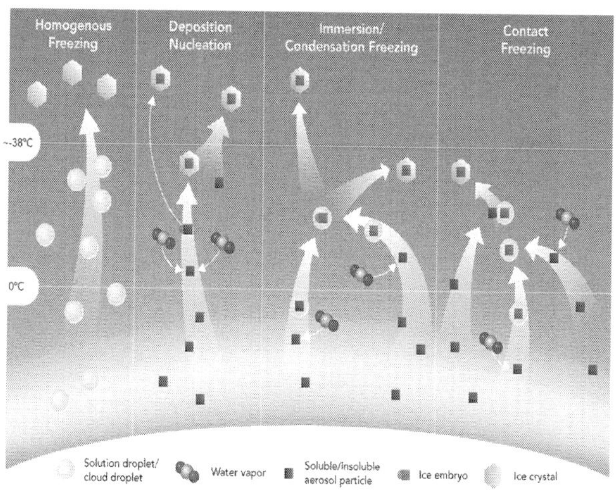


**Figure 5.6** Bergeron-Findeisen process.

**Source:** *http://pressbooks-dev.oer.hawaii.edu/atmo/chapter/chapter-7-precipitation-processes/*

Collision will result in the production of the water droplets and there will be the formation of warm clouds. If the ice crystals and moisture droplets are large, they can get deposited onto the ground. It is to be noted that the size of the crystals and droplets should be at least 0.1mm so that they can fall on the ground. Most of the raindrops that fall on the ground can be in the size range of 0.5 to 3mm. In the case of a drizzle, the size of the droplets will be <0.5mm. The moisture and clouds in the atmosphere will be dominating at 550m above sea level. The important mechanisms that can lift water vapor can be convective lifting, orographic lifting and frontal lifting (Vit., 1999). The basic theory that supports the surface runoff was proposed by Horton in 1933. According to this theory, if the rainfall rate is much higher than the infiltration capacity, we can define it as surface or overland runoff. This type of overland flow can be called Hortonian runoff or Hortonian overland flow (Figure 5.7).

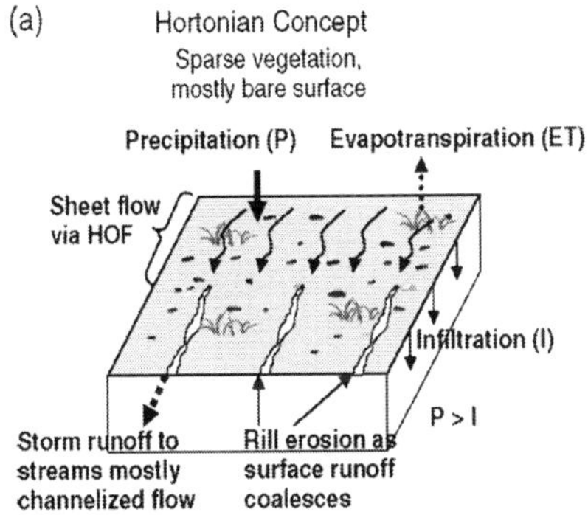


**Figure 5.7** Hortonian runoff.

**Source:** *https://www.semanticscholar.org/paper/Hortonian-overland-flow-from-Japanese-forest-the-or-Sidle-Hirano/b37a2ce-4a2e9e4c484f273254f2e071e1d04acb0/figure/0*

There are two important mechanisms that can be considered while studying hydrology studies. They are saturation excess and infiltration excess. There are many models that can be used to calculate the runoff that might have occurred due to rainfall events. These models will consider catchment scales. In the case of small catchments, the catchment area will readily respond to rainfall events. The response rate will be rapid. The NRCS method is a popular method that can describe two flow regimes that are sequential. They can be called shallow concentrated flow and sheet flow. In the case of sheet flow, the runoff can occur via continuous sheets (Water) over the surface of the land. There will be the formation of gullies and rills at large scales. Generally, the surface runoff can be through drainage pipes and open channels. We need to consider peak runoff models to measure the peak runoff rates. These peak flow rates are important in designing culverts and bridges. We can use these flow rates in the evaluation of the flood potential. The low impact developments must be encouraged to save the local ecological functions and hydrological functions.

We can use continuous-runoff models to measure the runoff caused due to rainfall events. We can use continuous run-off models to estimate the surface runoff. This runoff can be based on the excess rainfall that might exceed the local drainage capacity. We need to consider depression storage, infiltration and abstraction while applying this model. We can use infiltration models to know the distribution of the rainfall. The surface conditions mainly affect the local characteristics of the catchment area. These models can be classified into four types. They are non-linear reservoir models. Kinematic wave models, time-area models and unit-hydrograph models. Mostly, these models can be applied to know the peak run-off rates. Evapotranspiration can be classified as an important consumptive factor. This can be used to quantify the water budgets. Water readily gets transformed into the vapor phase from the liquid phase. This can be called the evaporation process. Water can get escaped from the plants via stomata. This process can be called transpiration. Most of the ground surface can be covered with vegetation. It is a tedious process to demarcate evaporation and transpiration. When we link evaporation and transpiration, we can get the value for evapotranspiration. This can be denoted as ET. We can classify the evapotranspiration into potential evapotranspiration, reference crop evapotranspiration and actual evapotranspiration (Abbott., 1987).

## 5.3 EVAPOTRANSPIRATION

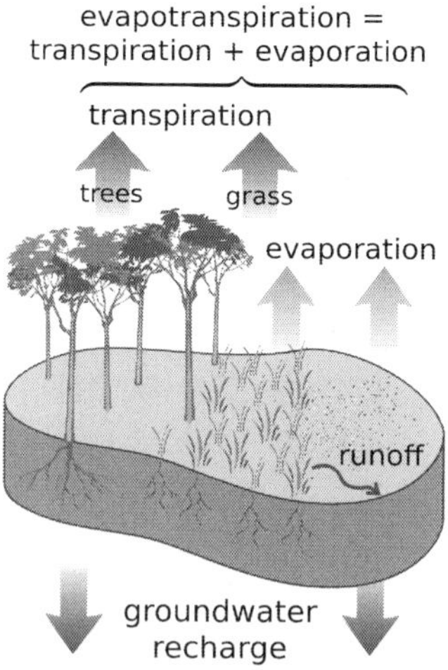


**Figure 5.8** Potential evapotranspiration.

**Source:** *https://en.wikipedia.org/wiki/Evapotranspiration#/media/File:Surface_water_cycle.svg*

The potential evapotranspiration (Figure 5.8) can be denoted by PET. It is nothing but potential evaporation. It defines water that is being evaporated at a location with time. This will be calculated under normal atmospheric conditions. Reference crop evapotranspiration can be defined as an ET rate from a crop. In this case, water availability may not be classified as an important limiting factor. This can be used to calculate ET. We can use this reference crop ET to estimate the evapotranspiration from the crops or vegetation surface (Bandoc., 2012). The actual evapotranspiration can be a form of evapotranspiration that can give a rough estimate of the water availability conditions, ground-cover and actual soil water. This can be called crop ET. The semi-empirical and empirical methods can be used to estimate the evapotranspiration rates. These methods can consider temperature, radiation and evaporation-pan methods. The temperature and radiation methods can be linked with ET with the air temperature and net radiation. The water bodies can be connected with the lifeline of humans. The maintenance of the river basins that were classified as transboundary can be considered a complicated topic. A river basin can be any area of the land parcel that holds all the surface run-off lows via rivers and streams. Some of the run-off lows can enter the deltas, estuaries and mouth of the rivers. Most of the river basins can be classified into elementary basins. These basins can be called sub-river basins. It is to be noted that the water management boundaries and basin boundaries may not be treated as compatible. The water supply systems that support agriculture and urban water supply systems may not follow via basin boundaries. The river basin can be declared and brought to attention for the zones that share the same boundaries. There are many prominent roles of the organizations that manage river basins. These roles are diverse. The organizations that are associated with the transboundary basins should take into consideration information, consultation, decision, implementation of programs and projects, resource protection and respective arbitration. There should be management and resolution and of the conflicts in the arbitration process. It is essential for pollution source inventories, alert systems and water quality monitoring in case of resource protection. The project evaluation and fieldwork can be considered in the case of the implementation of programs and projects.

Most of the river basins can be classified as social-ecological systems or SES. In the SES, we can consider a multitude of scales, sectors and players. Most of them are interdependent. Due to the inherent complexity in the SES, we can expect a high range of governance and management. The rapid development in technology, economic and social variations can enhance the complexity. Climate change can enhance the variability and uncertainty in rainfall patterns. There will be enhanced rainfall variability in case of climate change. This will also result in extreme weather events. It is always important to relate adaptive management and socio-ecological systems with cross-scale interactions and multi-level governance. There are two primary factors that can be used to attain transformation due to adaptive management practices. There should be political independence. The important individuals should exhibit exceptional leadership so that the change can be appropriate. This can be called transformational leadership. There should be actor groups, creative teams and framed creativity. It is important to uphold social memory and experience (Kircham., 2014).

The rivers can be classified as important watercourses that can flow naturally. Most of the rivers will hold freshwater. These rivers will meet another river, lake, sea and oceans. Rivers may flow into the ground. This can result in the drying of the rivers. These rivers may not reach the seas or any water bodies. There will be small rivers that can be called rill, rivulet, brook, creek and stream. In some places, the streams can be labeled according to their sizes. Mostly, rivers can be considered as an essential component of the hydrological cycle. Water can get collected in the rivers via precipitation. This can be via drainage basin and runoff over land. Some of the other sources that can feed rivers are springs, groundwater, snow packs and natural ice. The streams and rivers can be classified as important features in the landscape. It is to be noted that at least 0.1% of the land is covered by these rivers. Most of the human civilizations and cities can be based on freshwater from the streams and rivers. The study of rivers can be called potamology whereas limnology is the scientific study of the inland waters. The rivers can start from one source or multiple sources. The rivers will follow a certain path and it is destined to mouth or ends at the mouth. The water present in the river can be restricted to channels. This can be generated from the stream beds between the banks. In the case of rivers that are larger, there will be the formation of wide floodplains . These floodplains will be shaped by the river course. There will be over-topping of the channel by the floodwaters in some cases. Most of the floodplains can be wider and this can be associated with the size of the river channel. The difference between the floodplains and river channels may not be obvious.

The rivers can flow via gorges, canyons, plains, valleys and mountains. Upstream or upriver can be related to the flow that is against the regular flow gradient. The downstream or downriver can be associated with the flow along the direction of the natural gradient. Mostly, the river channels can comprise a single stream. In some cases, the river flow can be affected by interconnected streams. The water flow that is produced/formed due to these interconnecting streams can be called braided rivers. We can observe the braided rivers in New Zealand and South Island. We can observe braided rivers in the peneplains and river deltas also. The braided rivers share similarities with the anastomosing rivers. The sediments can be transported with the aid of multiple sinuous channels. We can observe the bifurcation of rivers also. The river bifurcation can be observed in Kosovo and Nerodime rivers. The flow of rivers can be in channels and this can be used as an energy source. The river channel can alter its shape. The weight of objects that are submerged in the rivers can be proportional to the $6^{th}$ of the speed of river flow. This law can be called Airy's law.

If the speed of the flow is doubled, then the river flow can transport or displace objects 64 times than the normal flow. All the objects that are submerged can be transported with the increased flow rates. In the case of torrential zones at the mountain terrain, we can observe erosional channels. These channels can pass through the hard rocks and hence there will be the production of gravels and sands. They may form due to the continuous weathering of the massive rocks. The glaciated valley that can be 'U' shape can be produced due to 'V' shaped channels in most of the regions. There will be meandering of rivers due to erosions that might have occurred in the river banks. In some cases, there will be deposition of the sand materials inside the bends. In rare cases, the rivers can cut a loop and hence channels can be shortened. There will be the formation of billabong and oxbow lakes. Most of the rivers can transport large sediment deposits. There will be the formation of deltas. These deltas can form near the mouths. In some cases, the mouth of the rivers can be saline. It is to be noted that some rivers may not flow under the surface. Some rivers may flow under the surface and they can be called subterranean rivers. These rivers can flow under caverns and caves. These rivers can be mainly observed in the geological formations that are made up of limestones. The subglacial streams can be classified as braided rivers. These rivers can flow at the beds of ice sheets and glaciers. The glaciers melt and the water gets discharged from the front portion of the glaciers. In some cases, the streams generated from the glaciers can flow uphill. This can be due to pressure gradients generated from the glaciers. The permanence of flow can be considered while describing rivers. Some rivers might flow rarely and they can be dried conditions for most of the tears. These rivers can be called ephemeral rivers or intermittent rivers. These rivers can be formed at locations where there is scanty rainfall. It can be assumed that these rivers can be prevalent at the river beds that are highly permeable. There are some ephemeral rivers that can only flow in the summer season. These rivers might be frozen in the winter seasons. These rivers can be called bournes.

## 5.4 RIVERS

Rivers can be classified based on several parameters. Some of them are topography and biotic status. The rivers can be mostly classified as bedrock and alluvial. In some cases, we can classify rivers based on two parameters combined also. The alluvial rivers can exhibit floodplains and channels. These rivers may form due to weakly consolidated sediments and unconsolidated sediments. These rivers can erode the banks and the materials can be deposited on the floodplains and bars. The river cuts the sediments and bedrocks; we call it bedrock rivers. The bedrock rivers can form in the regions that are previously uplifted. They can also form if the local lithology is hard. We can observe steepening river gradients in this case. The bedrock rivers can contain alluvium mostly. The alluvium is important in the sculpting and eroding processes of the channel. The mixed bedrock alluvial is the rivers that can pass through the patches of the deep alluvial cover patches and bedrock patches.

Mostly, these alluvial rivers can be classified as straight, wandering, braided and meandering channel patterns. The morphology of these alluvial rivers can be regulated by the combination of bed aggradation, vegetation, discharge, substrate composition and sediment supply. The top surface layer of the river can be called a crenon. The crenon can be classified into hypocretin and eucrenon. The eucrenon can be called a boiling zone or spring zone. The hypocretin can be classified into headstream zone and brook zone. Mostly, these areas can exhibit slow moving water, lowered oxygen and low temperatures. The upstream portion of the river system can be called a rhithron. The rhithron can be located below the crenon layer. This zone or portion can be demarcated by the fast flow of water, turbulent flow, high oxygen levels and cool temperatures. The downstream portion or stretch of the rivers can be classified as potamon. This region can be represented by sandier materials, slow water flow, low oxygen levels and there will be high temperatures.

## 5.5 STREAM ORDERS

The stream order can be designed based on the hierarchy and connectivity of the tributaries. It is to be noted that most of the rivers fall in the first and second order. The classification of rivers can be based on the stream order. There are some organisms that can survive in riparian zones. These organisms can be affected by the fluctuations in the flow patterns and location of the river channels. It is worth noting that the river ecosystems are river continuum sectors. There will be always certain additions and removals from the river ecosystems. The first-order streams can comprise decaying matter that might have derived from the forests. Plecoptera larvae can shred these decaying materials. Most of the rivers can flow downhill and this can be due to gravity. All the factors can contribute to the complex meandering process. The rivers that can flow downhill can move from the source to the mouth of the rivers. These rivers might not take the shortest route. In the case of alluvial streams, the braided rivers and straight rivers can exhibit limited sinuosity. The direction of flow can be downhill.

The rivers can be subjected to meandering effects. The rivers can flow sideways and this will be along the valleys and in some cases across the valleys. The bedrock rivers can flow in diverse patterns. Mostly these rivers exhibit fractal patterns. These patterns can be due to the low strength of the bedrocks. The limited strength of the bedrocks can be due to erodible layers, fractures and faults (Walsh., 1972). The volumetric flow rate can be called discharge and this will describe the rate of water flow. The flow rate of water can be studied at the cross-section of the channel of rivers with time. The open channel flow can be described as a flow (liquid) on the sloped channel and there will side walls and a solid bottom. The open channel flow can be divided into several types. They can be classified as unsteady and steady flow. In case the flow depth is constant with time, we call it steady open channel flow. We can go for greater depths, there will be a high level of unsteady flow. The open channels can be varied and uniform flow. If the flow rate is the same all along the channel, then we can call such an open-channel flow a uniform flow. In some cases, the flow can differ with the depth. Meander can be considered as a loop that might appear in the river channel.

The wetlands can be defined as those lands that can be saturated with water for prolonged times and they may support vegetation easily. These wetlands can regulate nutrient flow and also filter the sediments and pollutants from the rivers. If the surface of the water meets the banks, then we call such a stage of the river as "bank full discharge". We can call this a dominant discharge also. In case the river discharge enhances, water can overflow on the banks. This can result in spillovers. Then the lands that are adjacent to these spillovers can be called flood plains. The water level that is present in the river can be called its stage. This water level or stage will vary according to environmental factors. The flow resistance, bed slope and cross-section can vary according to the velocity and depth of the rivers. If we are studying a small portion of a river, we need to consider total energy, water surface and river bed variations. There are two important methods that can be used to estimate the surface profile of the rivers. They are graphical integration and direct integration methods.

## 5.6 FORECASTING

The forecasting of river flow can be considered an essential function of the planning process. This forecasting process can optimize the use of water resources, hydropower systems and irrigation systems. The forecasting process can help in the operational river management and this is because of the irrigation needs, industrial demands and population growth. The forecasting process can help us predict flood events and helps regulate the reservoirs. The data collected from the forecasting process can control the reservoir›s input and output. This is essential even in the decreased flow days. It is worth noting that floods kill many people annually all over the world. The prediction of the flood duration and magnitude can save many lives. Due to the advantages offered by the forecasting models, the forecast methods were classified to measure the river flows. The statistical methods can be applied in the hydrology fields. These methods are important in the forecasting of river flow and flood events. The statistical methods can be used to find out the relationships between data procured from various sources (Adams et al., 2016).

Out of several important forecasting methods, seasonal autoregressive integrated moving average or SARIMA and autoregressive integrated moving average or ARIMA method can be classified as popular methods. Artificial neural networks or ANN can be used instead of statistical methods in addressing several problems. These models can easily work with the nonlinear data and we can also use the flexibility offered by the artificial neural networks. The river flow can be represented and modeled using shallow water equations that are generally non-linear. The numerical predictions of the river flows can be proportionate with the depth of water flow. We can use depth-averaged equations in this case. It is important to know that the 2D shallow water equations are vital in the modeling of coastal and inland river systems. It is always essential to maintain a steady solution while calculating the water depth scenarios that might occur in the river systems. The balanced conditions should be ensured always. The bed slope values should be calculated properly. Mostly, the systems will be conserved.

It is always necessary to understand the anthropogenic effect on the river systems. The appropriation of water resources and rivers by humans can always yield stressed conditions in the ecosystems that are dominated by freshwaters. The ecosystems may demand certain human interventions to rejuvenate the damage caused by human-induced works. Exhaustive utilization of the river ecosystems can alter the natural flow regimes and this will result in damage to the biotic and abiotic components per se. The functioning and structure aspects of the river ecosystems can alter the water quality. Some of the ecosystem functions that might be altered due to human-induced stresses are cessation of organic matter breakdown, respiration and lowered gross primary productivity. Some of the biological communities like fish, invertebrates, algae and bacteria can be affected due to the anthropogenic pressures on the river ecosystems. Several microcontaminants and nutrients may be increasing or decreasing. The water stress conditions can lower the phosphate concentrations in the river environments. Most of the pharmaceutical products that are dumped into the local water will get increased. The harmful effects that damage the river ecosystems can be based on the size of rivers, water stress type, season, rainfall frequency and climatic conditions.

The usability of the water resources can be considered as an important characteristic to explain the cross-linkage between humans and nature. These manifestations might get increased with the land-use changes, climate change conditions and human population growth rates. Because of the overutilization of water resources, there will be water scarcity and structural deficit. This will directly affect the economy soon and at least 1.4 billion people will get affected due to water scarcity. There will be damage to human health conditions if water scarcity continues. The anthropogenic factors can effectively damage the riverine ecosystems that are dominant with freshwater. Due to the human-induced changes in the river ecosystem, we can expect problems in the quality and quantity of water. The quality of water can be represented with lowered biodiversity, pollution and eutrophication. In case of problems within the quantity parameters, we can expect flow regimes that are deeply altered due to overexploitation of water from the rivers.

In the water-scarce regions, it is always important to properly manage the watercourses. The industries, urban areas and agricultural sectors always demand water for their routine activities. There are certain practices that can supply water to the aforementioned sectors. They are groundwater exploitation, channelization, dams and weirs. If the water is being routed through these practices, then we can expect a change in the water quality and loss of biodiversity. The water stress that might occur naturally can be demarcated from human-induced water stress. We can expect cessation and flow reduction and predict in case of temporary rivers or we can call them intermittent rivers. These rivers exhibit decreased flow rates as time passes. Climate variability can also affect the river flow regimes and this can be long-standing. Some of the species that inhabit the intermittent rivers can get adapted to the change in climate and this can be called resilience or resistance towards climate change. The human-induced water stress can also generate alterations in the flow resulting in unprecedented flow dynamics. The unregulated flow can change the biological communities that are previously undisturbed. These biological communities may not be adapted to the latest disturbance caused upstream. In the case of semiarid and arid river ecosystems, human-induced problems can bring about artificial droughts. This can be called water flow intermittency caused by humans. In situations that are less than worse in the river ecosystems, we can expect certain levels of water stress by default. This can be due to natural riverine conditions that may exist in the local environments. The natural dynamism and synchrony of the river ecosystem will be reduced. The drying out conditions may prevail in some cases. Mostly, the water-stressed systems can be represented by minute peak flows that are not frequent and there will be reduced flow periods. The stability of the river hydrology can be affected and this will lead to a certain rate of natural dynamism.

Most of the anthropogenic activities can result in changing the course of the freshwater ecosystems as they are the sole resource of drinking water. It is to be noted that the concentration of pollutants and nutrients can exhibit certain patterns. These patterns can be attributed to sediment transport and stream habitats. The diversity of the biological species, their abundance and composition can get affected due to lowered peak flows. There will be an increase in the metabolic rate of the ecosystems and this can be due to excess production and survival of the primary producers in the river ecosystems. These primary producers will be dominant over the stream beds. The reduced flow rates can bring about varied responses in different river ecosystems. There are specific environmental parameters that can regulate rivers and this can be matched with the stresses that might occur in the river systems.

## 5.7 ARTIFICIAL INTELLIGENCE

Artificial intelligence (Figure 5.9) or AI techniques can be used in predictive modeling. We can easily obtain data from the natural systems that were classified as nonlinear. In the case of cold weather zones, we can expect a high rate of seasonal variability and this will be a tedious task in the forecasting process. This can be data demanding and complex. It is always a laborious task to model watersheds that occupy large areas. This can be attributed to the hydroclimatic fluctuations over the watershed under investigation. Some of the hydroclimatic factors that can alter the river's properties are soil drainage capacity, the density of basin drainage, land cover, climatic patterns and topographic conditions. There are several other factors that can be classified as hydroclimatic factors. The fluctuations in the river flow magnitude according to seasons can result in decreased water supply for domestic and industrial purposes. This also can affect the natural ecosystems that might depend on freshwater ecosystems. The ecosystems that survive at the streamside and some of the freshwater species that exist in the floodplains can be affected by fluctuating hydroclimatic conditions. These variations can also force the water resource professionals to review the water management plans. It has become a hectic task to alter the objectives of the reservoirs with the changing climatic conditions. It is vital to discuss the spatial and temporal variables before framing a working principle in the case of reservoirs. The study of hydrological regimes can be vital in water resource management programs. It can be important in building sustainable projects that can lower the damage to natural ecosystems. We can classify the forecast models of the river systems into two types. They are data-driven models and process-driven models (Corina et al., 2017).

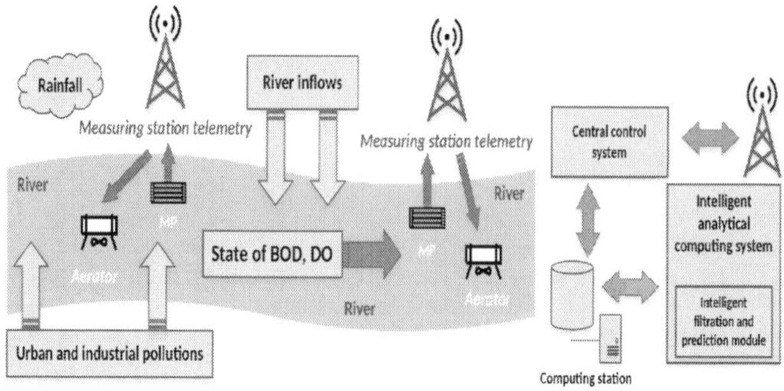


**Figure 5.9** Artificial Intelligence.

**Source:** *https://www.pinterest.ca/pin/774619204630612393/*

## 5.8 MODELS

The process-driven models can be used in the simulation of the physical processes using mathematical methods. These methods can be useful in the watershed boundaries only. In this method, we can combine both physical-based equations and empirical equations. We can consider data-driven models as nature's black box. These processes may not need deep knowledge of the river systems as most of the features will be simulated using the data provided by the agencies. We can use empirical equations to relate the field data followed by calibration. The data needs and the processes that guide these analyses can get different if we compare these two models. We can use software for these purposes as mostly open-source platforms are available. The topographic data, precipitation, and temperature data collected from the ground stations can be important. We need to use the archival data obtained from ground stations as valuable model inputs. The physical-based models can help us know the hydrological processes that can occur in the watersheds. We need several variables to calibrate the model. It should be known that the physical-based models are hard to achieve. It is always important to opt for real-time forecasting. This can be done using data-driven models and physical models that are advanced.

The study of the open channels can be linked with the shape of the river ecosystems. It is important to know the fluid dynamics of the river. In the case of hydrodynamics, we need to understand the fluid dynamics of the open channels. The fluid properties were studied by Greeks. Archimedes has proposed hydrostatistics. It was the Romans who applied fluid flow laws on the open-channel flows. During their tenure, there were water wheels and aqueducts. This equipment helped many farmers and the populace to satisfy their needs. Advanced siphon systems were used to seal pipes. This sealing process helped them to maintain sufficient pressure in the pipes. They clearly understood the concepts like fluid potential and pressure. The physicists and engineers of the 9$^{th}$ century applied hydraulics and fluid flow to perform several mechanical tasks. Most of the channels were engineered to build qanats in the irrigation process. The water wheels were used to generate energy. Mathematically speaking, the open channel flows did not get advancement in the case of their applications. The study of open channel flows is never an easy task. This difficulty can be attributed to the least estimation capability of river depths by us. The pipe flow can be easily studied as we know the cross-sectional area in advance. The problems with the calculation of certain estimates can be due to no accurate measurement of cross-sectional geometry, discharge, and depth. This complexity can be attributed to the varying bank materials and bed variations. Because of these complexities, we can frame empirical equations only.

Mostly, the river morphodynamics can be related to the shape and composition of the river beds/ The shape of the rivers can be studied to the geomorphological features. We can associate the morphodynamics of the rivers at variable spatial scales. We can consider meanders, large bars, and ripples. The generation of the geomorphological features can be linked with sediment deposition and entrainment. In the modeling process, we need to consider all the factors that control the sedimentation process. These factors can be studied on diverse spatial and temporal scales. We can scale up the possible effects that might be recorded at small spatio-temporal scales into large scales. Most of the processes that might occur at large scales can be studied to the boundary conditions. This can help us know the effects that might be more pronounced in the smaller scales. While studying the river networks at large spatial scales, we need to consider sub-basins and river basins. We need to consider water discharge, solid discharge, lake sedimentation, reservoir sedimentation and soil erosion whilst studying the river basin. We can perform basin-scale studies with the data collected from river drainage networks. In these studies, we can consider sub-basins and delta. We can use tools like one-dimensional, two-dimensional run-off models along with geographic information systems to know the river basin studies. Any river reach that has a longitudinal bed slope can be stated as morphodynamics equilibrium.

Most of the sedimentologists, geomorphologists and hydraulic engineers focus on the rivers as they carry solutes and sediments from the drainage areas. The river systems and the areas that are adjoining them can be affected due to sedimentation and erosion. The river ecosystems can change according to the loads that are being deposited on the beds. The loading rate can vary according to the flow patterns and magnitudes. Most of the engineers are worried about changes that have occurred in 20 years and some will even take into consideration 100 years of data. The change in the river channels can be based on the plan-form, slope, and material of the bed composition, shape, and size of the bed. It is our duty to understand the mechanisms that are regulating sediment deposition, transportation, and erosion. With the data collected, we can frame certain qualitative and quantitative methods that can portray the behavior of rivers. We can call these studies fluvial hydraulics. In some regions, this study was called river dynamics. The geomorphologists and geologists can consider qualitative methods in their studies. We need to consider the form and structure of the river ecosystems. Geomorphology can be called physiography in some areas. This study area includes mathematical geography, oceanography, meteorology, and climatology, etc. The action of the rivers can be called as the fluvial nature of that river. The science that can consider the action of rivers on the landforms can be called as fluvial morphology. In some areas, it is synonymous with river morphology. The systematic observation of natural phenomena has led to the development of the sciences like geomorphology and river morphology. These sciences can be called descriptive science. It is always essential to keep an equilibrium between the anthropogenic interferences and natural interferences as far as the streams are concerned.

If a large dam is installed over the river, it is important to store water that might be used in power generation, recreation, flood control, water supply and irrigation. The transport of sediments can be reduced by cleaning the dams periodically. Due to sedimentation, there will be flooding and depletion of the actual reservoir capacity. Water that is released from the reservoir can be classified as sediment-free. The sediments can be added to the flowing water from the banks and beds of the streams. This can result in the production of the extended reaches. This can also result in the widening of the channels. The degradation can also bring about several bad effects on the natural ecosystems. The transport of goods through waterways can be cheaper than rail transport and road transport. The extractors, sediment excluders, canal headworks and barrages can be used to withdraw water that is sediment-free. This water can be used for water supply and irrigation purposes. Because of this extraction, the equilibrium will be disturbed and this will yield aggradation in the reaches that are downstream. If the mining wastes are dumped in the rivers, then there will be aggradation in the rivers. The increased excavation of the gravel and sand from the river bed can damage problems in the reach. There will be degradation in the sub-tributaries and tributaries. There will be equitable distribution of water among the territories in a country. These distributions can alter the equilibrium conditions of the streams. There will be a balance between the distribution of the sediment loads and water distribution. This delicate balance will be disturbed. The construction of the floods can be done via straightening of channels, cut-off meanders, the building of reservoirs and embankments. The dredging process that might occur on a large scale can be used for navigation. This will also alter the stream equilibrium and sediment balance. The important factors that can alter the stream stability are alignment, cross-section, plan form and slope of the channels. Mostly, the urbanization process can bring about large-scale changes and this can be witnessed in the catchments. Most sediment yields and hydrological characteristics will be changed or minimized. Due to increased storm flow, runoff, usage of construction equipment, deforestation and mining of new grounds can enhance the land erosion rates. Once the urban areas are extensively developed, we can observe reduced infiltration rates. The groundwater levels will also get reduced. Due to rapid urbanization, there will be floodplain encroachment. The river channels can be restricted to extreme flood levels only. The prolonged variations in the hydrological and climatic regimes can yield varied discharge levels. The nature of the sediment load can affect the channel dimensions and quality. This can lead to an altered river course. In some cases, the rivers can get perished. The tectonic movements and earthquakes can affect the stability of the rivers.

Chapter

6

# River Hydrology

## CONTENTS

6.1 Introduction....................92

6.2 Hydraulics....................94

6.3 Floodplains....................97

6.4 Braided Rivers....................98

6.5 Watershed....................105

6.6 Sediment Transportation....................110

## 6.1 INTRODUCTION

The rivers can take away certain ingredients of the earth and it can be witnessed through dissolved materials and erosions that might be present on the earth's surface. It is often a hectic task to demarcate the dispatched materials from the earth's surface due to anthropogenic activity and river-induced activities. There are many rivers that are affected by the sedimentation process globally. This will reduce water use by the industries and agriculture sectors. The ecological integrity will be diminished due to the sedimentation process. The rivers can generate excellent picturesque patterns for the spectators and provide essential drinking water to the public. The eutrophication by the nutrients and sedimentation can threaten the basic characteristics of the rivers. It is important to know that the agricultural rivers can drain the soils and erode them. The reduction of the sedimentation process is important to rejuvenate water quality and uphold biological integrity. The rivers can be classified as arteries of the earth and they can easily drain water from the surface to the oceans and seas. Water can carry dissolved materials and sediment particles from terrestrial environments. It is obvious that remote sensing technology has provided us with a lot of information about rivers and their essential characteristics. Due to the availability of aerial imagery at high resolutions, we can understand the boundary conditions and fluxes of the river systems. The reduction of sediments may need a complete understanding of the erosion process and sediment sources. The forests, wetlands and other land cover transformations can bring about visible changes in the river ecosystems. The stream flows can affect the habitat, morphology of channel, transport of nutrients, sedimentation, and water quality. Some of the features of stream flow that are to be considered are timing, duration, frequency, and magnitude (Montanari., 2012).

It is important to know that the rocks can be composed of ancient formations made up of minerals, vegetables, animals, and bodies. Most of the rocks are being subjected to decay and these materials can get deposited into the massive river bodies. The morphological processes may need longer geological time scales. The relative change that might occur in the sea levels and land can be considered as an important factor. The drainage basin can be defined by the tributaries and streams. The drainage basin can be called a catchment area or watershed. The drainage patterns can be associated with the arrangement of different streams in the basin regions. Mostly the drainage patterns can be described as the regular arrangement of the channels at some drainage basins. The drainage patterns can affect geomorphology, diastrophism, structural controls, and the hardness of rock.

Mostly, these drainage patterns can be controlled by several factors. We can classify drainage patterns into many types. In some areas, we can observe branch-like and dendritic patterns. In these patterns, we can notice irregular branching of the streams and this will be diverse. The angle of deviation of branching can be less than $90^0$. We can observe these patterns in the igneous rocks that are massive and sedimentary rocks too. We can observe these patterns in the metamorphosed rocks also. In some regions, we can notice lattice-like or trellised patterns. We can find sub-parallel streams also. These patterns can be noticed at the stroke of the rock formations. These patterns can be parallel with the topographic features. These features might be deposited by ice or wind. We can observe radial patterns on the volcanoes and flanks of the domes. The radial patterns can be observed in the sub conical hills and isolated conical hills. We can observe parallel drainage at the slopes. There will be structural controls that can yield spacing between the parallel streams. The rectangular drainage can be seen in the main streams. The tributaries can exhibit bends that are right-angled. We can notice faults and joints arranged at right angles. These streams can be modified according to the structure underneath. The deranged drainage patterns can be observed if there is no bedrock control or structural controls. Preglacial drainage can occur due to the glaciation process. Some of the irregular streams can enter the swamps and lakes. The centripetal pattern can be observed in the local areas. Most of the drainage lines may get united with the central depression. These can be noticed in the basins with depressions, craters, and sinkholes. In the case of complex geological conditions, we can observe patterns that are highly violent (Biao., 2017).

There will be certain drainage patterns that can be noticed. We can expect geological activities, faults, joints, slopes and lithology with limestone solution, volcanism, and glaciation. The texture of drainage can be considered as one of the essential features of the geomorphological nature of the drainage. We can classify drainage textures into coarse, medium, and fine. The climatic conditions can alter the nature of the drainage texture. The magnitude of precipitation can directly affect the nature of runoff. The topography with improper land mass can initiate drainage structures. The thundershowers, vegetation, shales, and clays can yield drainage structures that are finer. In the case of gravel and sand outwash plains, we can notice a coarse drainage structure. We can expect less drainage lines in the case of gravel plains. The structure of drainage can be considered as an important factor in determining the density of the drainage. The river net can be defined as a drainage pattern that can be interrelated with the stream sets in a specified area. These channels can meet at a point and that point can be called a junction. The stretch of a river that is unbroken can be called as a link and it can exist in between 2 junctions. We can define this as an interior link. If this link exists between the first junction and source, we can call it an exterior link. If the headwater streams are devoid of tributaries, we can classify them as first-order streams. If two streams get united, then we can classify it as a second-order stream. These second-order streams can get united to produce third-order streams. The basin area can be important and it can affect the hydrograph size of the storm. This basin area can also affect the peak flows and mean flows.

## 6.2 HYDRAULICS

The alluvial streams can be any flow that can carry sandy particles. These alluvial streams can shape the channels via the transport of sediments and water. We need to consider critical velocity and critical shear-related methodologies to understand the motion parameters of the river flow. It is important to know the modes that support sediment transport. It is mandatory to understand the bed forms. The flow resistance should be considered. The sediment transport can be studied in terms of total load, suspended load, and bed loads. Shear stress can act on the bed and sometimes this value can exceed threshold levels. Most of the sediments can be in motion in a particular direction. The flow of sediments can be based on particle size, fluid and sediment densities and flow conditions. In case of contact load, the sediment particles can slide and roll with the bed for a certain time interval. After a certain pause, these sediment particles can slide and roll. This can be called as contact load. This is nothing but the slide or material that is being rolled on the bed. This will enhance the contact times. The sediment particles can bounce or hop with the bed materials. Hence, the contact time will get reduced. This can be defined as saltation load (Pyron et al., 2011).

Saltation load is nothing but the bouncing of the sediments that may travel along the beds. Sometimes, the sediments can move indirectly or directly due to the impact of the incoming particles that are continuously bouncing. The demarcation of sediment load is not obvious so it will be integrated with the contact loads and sediment. This can be called a bed-load. Suspended load is nothing but any material that is being transported as a suspension in river fluids. Most of the suspension will get altered with the turbulent fluctuations. It is important to know the demarcation between wash load and bed material load. The alluvial stream can carry sediment load. This can be classified as wash load and bed material load. The sediment load that is being produced from the stream banks and bed can be called as bed material load. The sediments can be found in the stream banks and beds. Bed materials can represent the hydraulic conditions of the streams. The sediment loads can comprise several materials that are finer in size and hence we may not find these sediment particles in stream banks and beds. This can be called a wash load. The wash load can be deposited into the streams from the catchment areas. The sediment loads that are being brought into the streams can help us assess the native hydrological conditions of the rivers. It is a tedious task to know the quantum of wash load that is being carried by the streams. We need to notice that the wash load has certain size limits. In the case of flat slopes and sandy streams, the wash loads may comprise silt range and clay. These sandy streams can exhibit slopes that are flat. The streams in the mountain regions can exhibit steep slopes. In these regions, the wash loads can constitute fine sand and coarse sand. The lag effect can be based on percolation rates, the inertia of the sediments, gravity forces, suspension rates and fluid friction.

The transverse ribs can be defined as an arrangement of pebbles that are regularly spaced. It can comprise boulders and cobbles. They can be arranged at right angles to the flow of the river systems. The antidunes can be water waves and sand that can be symmetrical. The antidunes can move against the streams. Once these waves travel against the flow, there will be increased amplitudes. The steepness of water waves will be increased. This will be based on the sediment size, depth and velocity. Most of the attempts were made to know the accuracy of the computational methods that are linked with the estimation of transport rates of the load. This transportation rate was studied in the alluvial streams. Interpretation of the results is difficult as most of the investigators get the data from different sources. The data will be procured from several data points and by different professionals. The channels that might flow through the Sandy materials can be used to transport water for irrigation purposes. These channels can flow through banks and non-cohesive beds. There will be a limited rate of cohesion that might appear in the channels. These channels might involve discharging the materials at constant rates. Mostly the estimated sediment loads can be carried through his channels. In the case of a plan form, there will be Limited variations. Most British engineers devised certain methods to build these channels in the Indian subcontinent. With extensive analysis, there was a successful design for Channels that might carry sediment load and discharge that is known. The hydraulic geometry to alluvial streams can be important while studying river morphology and related studies. We know that most of the channels will maintain stable relationships with the adjoining water bodies. We need to take into consideration the slope, area, depth and width of the alluvial streams.

We can take into consideration hypothetical flow patterns. It can be called a dominant discharge or bankfull discharge. In some areas, it is being called as mean annual discharge. All these factors are important in shaping the channels. There are several Optimisation techniques that can be used to define the hydraulic geometry of the streams and channels. We can retrieve the relationships that might occur in hydraulic geometry using dimensional analysis methods. The flow that might occur in the alluvial channel bends and the flow patterns that are observed in the rigid belts must be considered to know the characteristics of the river ecosystems.

It is always essential to understand the flow while stream meandering takes place. We need to know the important characteristics of flow in the rigid belts and bends. We also need to consider the sides that constitute the alluvial boundaries. We need to know the distribution of velocities in radial and longitudinal directions. As for the secondary circulation is concerned, we need to estimate the decay and growth also. It is essential to know the bed topography, shear distribution, head loss specific to bends and superelevation so as to understand the actual characteristics of the flow patterns. The meandering and braiding can be studied in the case of river ecosystems. We need to understand and estimate the extent of braiding, parameters that affect this process, streams and bars along with the causes of braiding. There are certain stable channels that can participate in sediment transport. In the late 19th century, British engineers estimated V Channel dimensions and flow velocities in the river systems of India. The considered non depositing segments and non-silting segments of the channels were defined as Alluvial channels. These channels carried sediment loads and respective discharges. The floor was observed as non-cohesive. It was sandy material. Irrigation channels or the canals can be generated from the head works. At the head works there will be the elimination of sediments. This is to ensure that there will be a minimal amount of bed material load. It can also flow from the cohesive banks and Sandy beds. In some cases, these canals flow through banks and beds. In this case, there will be limited perimeter with greater depths. The greatest depth can be correlated with hydraulic radius. There will be smaller areas also.

We can use hypothetical discharge studies to know the hydraulic geometry that might occur in the streams. These studies were done in the USA, Europe and India. There is vast variation in the sediment loads and discharge that are being transported through streams and channels. The maximum and minimum values will show a high rate of variance. Most of the canals that are specified as hand canals can carry constant discharge. This discharge can exhibit higher sediment load variations. There will be high variability rates on bank material bed size. These variabilities are common in the streams. We can notice a wide range of standard deviations if we consider bed materials. A stable canal may present less than regular shape and irregular river cross-sections. A stable canal can have a plan form that can be fixed but in the case of streams, the plan form may vary with length. Features of the bed material and slope of the river main change continuously. There are several limitations in the estimation of dimensions of river channels. Deposition and erosion can be considered as driving factors that make these estimations hard to achieve. it is always important to understand the geometric features of the cross-sections that might occur in the river environment. These calculations can be associated with the constant discharge that is being assumed. We need to consider characteristic discharges and Bankfull discharges so as to reach meaningful conclusions. The constant discharge is a concept developed by several researchers to explain river systems and their flow characteristics.

## 6.3 FLOODPLAINS

Williams in 1978 has explained the advantages and disadvantages of different systems that described the bankfull stages. He demarcated active floodplains and inactive floodplains. The active floodplains were described as the spread of water annually during active floods that yield sediments. The inactive floodplains can be called terraces. It is nothing but I Valley flat that might be submerged during occasional floods. In the case of inactive floodplains, we might not expect the deposition of sediments. We need to consider the average elevation presented by the active floodplains. The water discharge and elevation ratio must be taken into consideration. In some cases, the ratio might be minimum. The bankfull stage must be determined to discharge and later we can rate the curves with the help of data collected from the gauging stations (Macdonald et al., 2011).

We can expect head loss when certain conditions persist in the local areas that support water resources especially reverse systems. There will be a loss of energy if the short bends are being separated. This can also be observed if the change in velocity is located vertically. The enhanced loss of energy due to internal friction mainly results in secondary circulation thereby resulting in a head loss. The boundary friction can also increase as this can be attributed to circulation patterns. This is also one of the reasons behind head loss. Most of the energy can be consumed in the secondary circulation thereby initiating head loss. The distribution of change velocity can be considered as a component of longitudinal velocity that matters along the stream width. The formation of secondary circulation can result in alteration of radial shear stress of the bed. It is to be noted that the altered shear stress can come with a minimal radial component. This can lead to the formation of cross-sectional beds. This deformation can be observed in the case of alluvial channel bends. If a circular Alluvial bend is a uniform and study, we can expect minimal sediment transport. This transport can follow radial direction and it gets counterbalanced by radial bed slope. This can yield greater depth at the walls.

The transverse bed slope can balance the sediment transport. Most of the river bends can be classified as meandering surface bends for entrenched bends. The entrenched bends can be found in the valley bends. The river that is present on the valley floor can be subjected to meandering surface bends and it is prone to erodibility or erosion. The characteristic of bank material can describe the curvature of the bend and its radius. These bends can be divided into force bends, Limited and free bends. In the case of free bend, we can expect alluvial material that can be subjected to erosion easily. In case of Limited bends, there will be e parent material that is consolidated thereby there will be Limited lateral erosion. In case of force bends, the streams that are under consideration will impinge on the bank starting from 60 degrees to 100 degrees.

## 6.4 BRAIDED RIVERS

The braided rivers (Figure 6.1) can be characterized by the presence of several alluvial channels. These channels will contain Islands and bars. This occurs in the Confluence of the rivers and the division of the rivers. There will be intertwining effects within this braided river reach. There will be a series of channel segments in the case of braided rivers. There will be Regular Re-joining and division in the braided river reach. It is to be noted that there will be a single dominant channel that regulates the braided river reach. It is important to us to know the causes of braided river reach formation (Nicola., 2015). There are different variables that can affect the braiding of rivers. Some of them are size distribution of sediment loads, the rate at which segment gets loaded, nature of bad material, geological factors, climate, slope, depth bandwidth.

**Figure 6.1** Braided rivers.

**Source:** *https://upload.wikimedia.org/wikipedia/commons/6/68/White_River_5965.JPG*

There are several rivers that can form channels. These channels can change their attributes within a short distance. The braided rivers may change into meandering rivers. These variations can be based on Independent variables that might occur locally. There will be high variability in the monthly discharge, total discharge and flood magnitude in the case of meandering channels. There are two main reasons for the formation of braided rivers. They are overloading and steep slopes. These streams may carry sediments and they get deposited in the streams themselves. In the case of steep slopes, we can expect Shallow streams and there will be the formation of islands and bars. Some of the important characteristics of braided channels are higher bed load, course or sand bed material, Shallow depth, wide channels, longitudinal slopes that are steep, straight course of the channel, etc. There will be the formation of longitudinal bars that are diamond-shaped. They are parallel to the direction of flow. They are tagged to react to channels preferably on both sides. They will be having eroded margins partially. Bars can be formed in gravel and it is considered a common type. The longitudinal bars can be formed by a multitude of factors sequence of events. In some cases, the length of the bars can reach hundreds of meters. We can observe horizontal bedding that is massive and, in some cases, screwed. Transverse bars can occur in braided rivers. Transverse bars can be called as languid bars. Mostly there can be classified as sand braided rivers. They are formed in the channels that are narrow and deep but restricted to narrow banks. We can observe ripples and dunes in this case. The lateral bars, side bars and point bars of similar if we consider their geometry. These bars confirm the areas that are interior to the meanders. The point bars can be linked with meandering rivers but they can also perform in the braided rivers. We can observe longitudinal deposition at the adjoining side of the rivers and so they are called side bars.

The important characteristics of the river that is subjected to meandering can be the mobility of meander loops, size and shape of the loops, etc. These characteristics play an important role in the maintenance of hydraulic structures like guide bunds, flood embankments, barrages and bridges. We can describe the shape of meanders as a parabolic curve for sine curve or in some cases it appears like a circle or arc. It is a well-known fact that the rotation of the earth metals meander in rivers. The surface of the earth can be subjected to Coriolis focus that might bring about inertia and water bodies and hence meandering occurs. It is to be noted that anybody that moves in the north to south direction in the Northern hemisphere may get deflected by the forces that are being generated from the eastern direction. The eroding nature of the banks can be important in describing the monitoring process. Rotation of earth can play a minimal role in the meandering process.

Disturbance theory is another popular theory proposed to understand the flow characteristics of rivers. According to this theory, reading is a service that is caused on the floor or on the bed. The upstream can definitely change the flow pattern of the river course downstream and that can lead to a meandering process. This disturbance causes variables to position along the channel width in a stream that is overloaded and there will be certain oscillations that can be observed in the course of flow.

It is always important to analyze the stability and frame criteria for plan forms. It is to be known that meandering can occur due to bed erosion that may lead to the production of channels with vertical walls. The gravel bed rivers can be called Boulder bed rivers. These rivers can flow through wire borders and gravels continuously. These rivers can be demarcated with sand bed rivers in having special characteristics. These rivers can comprise materials that are coarse than the sand bed rivers. We can expect steeper slopes in these rivers than the sand bed rivers. Almost all the fractions of the material constituting bed in the sand made rivers can transport in the form of discharge along with the minute particles. Gravel bed rivers can also transport all the fractions of the material that constitute bed but the floor will be Limited. Most of the sediment transport may be on the pavement regions as there will be occasional rainfall at those locations.

The structure of bed forms can be different in the sand bed rivers and the gravel bed rivers. There will be no formation of dunes and ripples in the case of gravel bed rivers and that is not the case with sand bed rivers as they conform to dunes and ripples. In the case of gravel bed rivers there II will be an e formation of bed features that are large. The features that constitute bed can offer resistance to the flow pattern and they also act as store houses for sediments. The slope of the gravel bed rivers can differ proportionately. This battery can be braiding and meandering according to the fluctuations of discharge in the river systems. The gravel bed rivers can act as transitional bodies or braided bodies. These rivers will exhibit a higher rate of tenacity than the sand bed rivers. The gravel bed rivers or generally much closer to the sources that deliver sediments whereas these and bed rivers are not that much too close to the sources of sediments. There will be great availability in the transport of sediments in case of gravel bed rivers. The variability is much less in the case of sand bed rivers. The transport of sediments is not uniform and study in the case of gravel bed rivers if you consider the water discharge that is studied. Based on the sedimentation supply to the concerned channels there might be variation in the transport rates for a discharge under investigation.

It is essential to understand the pavement characteristics and the river ecosystem. Most of the river beds can be constituted of heterogeneous mixture for small particles of gravel and other materials. These materials form a surface layer and it has gradually increased or decreased with depth. In certain times there will be no supply of segments from the upper portion of the river, that best surface is made up of particles the cursor. These particles will gradually become mobile if the rate of discharge is much reduced than the standard for conditions. If the flow rate is much higher then there will be higher mobility in the bed particles and hence the bed particles or the bed will gradually degrade and become coarser. These conditions can be witnessed in the lower portion of the Dam and also in cases that are mobile. The surface layer can be put in a class of gravel bed rivers and it is known as pavement. This can be maintained by continuous transport of bed load. This transport will be based on the size of the particles. Most of the sediments will be injected from the upper side of the stream so as to maintain s equilibrium in the channels.

The transport of sediments may occur continuously for several years. The coarse particles that are present on the bed can rarely move and hence they may not be considered in determining the stability of the bed. We need to understand the hydraulic geometry of the rivers with the help of hydraulic radius, depth of flow and average perimeter values. It is essential to understand the hydraulic diameter of the rivers by taking into consideration the slopes and Bankfull discharge. Dress of gravel bed rivers, the formation of meso scale and microscale dunes and ripples is scarce.

In the case of micro forms, the important features are lineation and ripples and this is quite absent in gravel bed rivers. In the case of meso forms there will be the formation of dunes. The formation of dunes is quite rare in gravel bed rivers, odd even absent in some cases. The Macro forms can produce antidunes, channel bars and unit bars. The formation of these components is rare but might be present in gravel ribs. In the case of Mega forms, there will be the formation of bar assemblages and hence the production of sedimentation zones with a high rate of variability. If a portion of the channel carries sediments then we can expect that the shear stress is drastically reduced. This is due to the fact that the bed load can get transported Into the Other zones of the river resulting in bars. Most of these bars can be formed at the starting points of the channel beds. We can observe this in the case of convex banks and wide channels where divergence is common. There will be the formation of spool bars or longitudinal bars in the river ecosystems.

The storage volume of reservoirs was estimated to decrease at least by one percent annually. This can be attributed to the sedimentation process that occurs due to improper management of reservoirs. This can be created with the volume that is being stored by constructing new dams. In order to protect communities from floods and to provide water supply, it is always essential to manage appropriate storage volumes and precious water resources. Most of the authorities and operators are involved in framing accurate sediment management practices so as to ensure sufficient storage volumes. 2D numerical modeling has been attempted in this sediment flushing process. The river sediment can be considered as a major water pollutant by volume and weight. These sediments can also work as catalysts, storage agents and carriers of a wide variety of pollutants. most of the river environments will get polluted if the sediments are not properly treated and flushed. if the sedimentation concentration is much higher then we can expect limited water quality. These segments can degrade water that can be otherwise used by aquatic life, hydrolytic facilities, cooling of Industrial effluents, recreation and also urban supply systems. Some of the wastes and chemicals will get adsorbed onto the sediment particles that are present in both rivers and reservoirs. There will be iron exchange between the sediments and solutes and this is common in most of the reservoirs that store water for longer periods.

Mostly, there is no water containing antifreeze agents, anti-corrosion agents, grease, transmission oils, lubricating oils, fossil fuels, polycyclic aromatic Hydrocarbons, organometallic species and heavy metals. Most of these chemical pollutants can severely affect the native chemical nature of lakes and rivers. This will be detrimental to aquatic ecosystems and biological species. There is much development in the agricultural and industrial sectors and this has led to enhanced pollution by heavy metals and enhanced formation of hazards that can damage humans, fish and invertebrates. Most of the heavy metals that are being produced by the industries are deliberately released into the rivers thereby creating a toxic environment for the fish populations. The sediments that are present in lakes and rivers play a vital role in determining the quality of water that can be otherwise used by human beings for drinking and various purposes. Most of the suspended sediments are capable of absorbing pollutants that water carries and hence the concentration of pollutants in the running water can be reduced to some extent. However, the sediments in the reservoirs are still toxic due to the accumulation of heavy metals on the surface. There are some pollutants that can damage the sediments by disintegrating a close pack of particles. Most of the benthic sediments can provide shelter to the fauna surviving in the reservoirs. The benthic sediments provide food to several species that are living in the reservoirs. Most of the pollutants may be classified as chemicals that can indirectly and directly be toxic to the fauna and flora of the aquatic systems. The effects of this pollution can also be witnessed on the terrestrial surface and this can be due to bioconcentration of pollutants in the food web and also bioaccumulation (Figure 6.2) or biomagnification.

Reservoir sedimentation is an important aspect of the Reservoir engineering process. It is seldom possible to differentiate rivers with negligible human interference. We can use the Reservoir index classification of reservoirs. If the Reservoir index is less than 10 percent then we call it natural. If the Reservoir index is between 10 and 50% then we call it semi-natural. The reservoir index is between 50 to hundred percent then we call it semi-controlled. Every reservoir index is greater than a hundred percent then we call it test controlled. The reservoir index is nothing but the ratio of all the storage capacity that is presented by the river reservoirs, the mean of annual runoff of the rivers at a specific location. The construction of reservoirs and their proper management can reap several benefits to society.

There will be several environmental benefits, recreational advantages, navigation, irrigation, power generation, water supply and flood control. In some cases, due to natural calamities there will be certain disastrous effects brought about by the reservoirs.

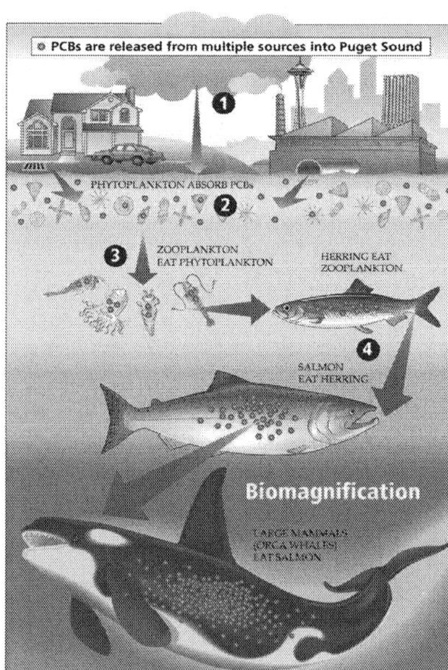


**Figure 6.2** Bioaccumulation.

**Source:** *https://cimioutdoored.org/bioaccumulation-and-biomagnification-increasingly-concentrated-problems/*

Once the reservoir starts storing water most of the sedimentation processes that occur in the reservoir will get interrupted. The sediments coming from outside or upstream can be Trapped inThe Reservoir zones. There will be the formation of a delta. Mostly the backwater can interrupt the sedimentation process. The sedimentation phenomena occur in the reservoir and hence lowers the capacity of the reservoir.

Sediments can be considered as a mixture of inorganic and organic materials that are being transported by rivers from one place to another place. This transportation process can demand energy. Due to sedimentation, the depth of the rivers will be reduced and hence the native structure of the pools and lakes will be altered. The sediment loads can define erosion of basin in, drainage basin, groundwater and morphodynamics. River systems have been a lifeline for many centuries to humans especially those who inhabit the coastal zones and river ecosystems. It is always important to judicially utilize the hydrological systems. The sedimentation in the dams and reservoirs has become a huge problem and water resource engineering practices. The deposition and sedimentation process can result in a meandering effect on the river ecosystems. It is always essential to frame certain sustainable management practices with both private and public government agencies. The management and estimation of sediments in a river environment are always essential. Because of the higher sediment loads most of the reservoirs fail well before their expected lifespan. Deposition and sedimentation can be considered as Natural river behavior mechanisms. This sediment load can exceed its limitations if not managed properly. In case of fine soil particle sediments, we can expect flocculation in the river systems. These fine sand particles can get suspended in the water bodies. Mud can be distributed all along with the mouth of the river and this can be enclosed by bulk statements. The ecological and morphological conditions can be disturbing due to sedimentation problems. We need to investigate the sedimentation problem using one-dimensional flood plains and related studies. Most of the segments that are being carried by rivers can change the pattern of the landscape and the physiography of the landscape.

## 6.5 WATERSHED

A Watershed can be defined as an area preferably land that allows all the precipitation deposited inter particular point like Lake, river or drains (Kassie., 2020). The watershed that is small can generally drain water to a larger watershed and finally, excess water will reach massive water bodies like oceans. The watersheds can effectively support wildlife and plant life that can be used for recreational activities. It is always important to regulate the watersheds using best management practices. The storm water runoff can be produced or generated from melting snow or rainfall. The storm water generally flows through storm drains, lawns, bare soils, parking lots, sidewalks, paved streets and rooftops. Whilst flowing, most of the run of will collect pollutants, litter, oils, fertilizers, pesticides and soils from the landscapes. Most of these waste materials will reach streams and rivers if not properly treated in sewage plants.

The strong water that is polluted can effectively contaminate the local waterways in many ways. The pollutants stormwater can damage the water quality and hence wildlife, fish and plants will die. Sediment is nothing but a loose collection of silt, clay and sand along with several soil particles that may settle down at the river bottom or at the base of any water body under investigation.

Most of the sediments can come from soil erosion and the decomposition of animals and plants. The environmental protection agency classified the common pollutants that contaminate reservoirs, lakes, streams and rivers. Generally, natural erosion constitutes or contributes 30% of the sedimentation proportion. 70% of erosion will be accelerated due to anthropogenic activities. The sediments that are concentrated can be produced from construction activities and home building projects. Sediments that added this from water can completely degrade the water quality and hence then local natural resources will get affected. Most of the sediments get deposited in the catch basins and storm drains. In general conditions, the catch basins and storm drains will be used to carry water that is left over by domestic activities and industrial activities. Mostly, water that is polluted with sediments can appear cloudy and so the animals may not recognize the food materials. Murky water can damage the natural vegetation. In some cases, we can observe a high rate of sedimentation at the beds of the streams. This results in Habitat loss and hence the total fish population will be reduced. The treatment of water will become harder if sediments are high. This will drastically increase the cost of treating drinking water so as to eliminate taste and odor problems. Some of the sediments formed can also obstruct the gills of fish resulting in reduced respiration potential of the fishes. This will lead to damage to the fish ecosystem. Most of the nutrients can be transported through sediments and they can activate the growth of blue-green algae. The blue-green algae can release toxins and the water that is dominated by this alga may not be useful. Some of the sediments that are deposited in rivers can change the water flow with reduced water depth.

The sedimentation process that occurs in the reservoirs can be classified into three fundamental categories. They are deposition of fine sediments, why are stratified flow, deposition of fine sediments with the homogeneous flow and deltaic deposition of course materials. Most of the reservoir's main density stratification develops because of turbidity variations, salinity and temperature. These factors may vary with layers of the reservoirs. Out of all these factors, the density stratification can play an important role in the circulation of sediments and water. The inflow of the river will be affected by the density stratification. The sedimentation process and erosion can be considered as a portion of the geological evolution of a particular landscape under investigation. Erosion on the surface of the earth can occur through water, wind and other methods. Deposition, transport and entrainment of the naturally occurring materials can be witnessed at any location provided the water source is intact. The geological time has shaped the local landscapes, intercostal planes, deltas, plains, floodplains, river valleys, and mountain gorges. The starting point of the aeration process will be due to the impact of Rain drops. The Rock and mineral particles will get detached and get migrated to shorter distance possible. Some of these particles may be carried by flowing water to Greater distances. The sediment grains can move from the mountainous regions to massive water bodies like oceans. The sediment grains are particles that can be temporarily deposited on a flood plain if there is a flood. The landscape evolution can be based on river processes and the river itself. Most of the rivers can be treated as transportation systems of sediment and water.

Net storage space of the reservoir has decreased in recent years due to irregular sediment management. Due to reduced storage volumes in the reservoir, there will be no extra storage capacity. In some cases, the flood level that was increased upstream can change the slope of the river. Dams and reservoirs are meant to control flood situations. The Reduction in storage space main demand revised operational Strategies and regulations so as to reduce risk due to flood situations.

There will be shifting of river banks and erosion resulting in the incision of beds in the downstream portions of the river. The coastal lines will also get eroded as the sediment supply through rivers is limited. The agricultural activities will be damaged or interrupted due to reduced supply of nutrients and fertile silt. This will also reduce the productivity of aquaculture like fisheries. The dynamic loads and static loads of reservoir structure will get started due to large sediment depositions before the spillway and Dam. Turbines in the hydroelectric plants will also get affected due to erosion. There will be clogging of hydro-mechanical equipment due to erosion. The vents, sluice outlets and flow control gates can get clogged. Concrete structures can be subjected to Cavitation and abrasion. Some of the concrete structures that can be affected by erosion are channels, bypass tunnels, cutoff walls, roller buckets and spillways. most of the aquatic ecosystems will be affected due to altered sediment quality. This can result in eutrophication of the reservoirs. It is essential to opt for randomized sedimentation removal practices due to large turbid situations that can persist in the reservoirs. This will lead to sediment hazards in the downstream river sections. Some of the main causes of sediment induced problems are deforestation and land use variations. Misappropriation of sediment transport and river morphology in the selection of design and locating appropriate areas can lead to problems with the Reservoir structure. Ignoring the morphology related issues can also bring about negative impacts on the reservoir and the population downstream of the rivers. The knowledge gaps integrated Reservoir management approach can also lead to disasters. The aging factor of reservoir sand dams must be taken into consideration. Improper maintenance of reservoirs can also lead to enhanced sediments.

There are certain parameters and processes that are to be measured while designing the reservoir and its operation. We need to measure groundwater, snowfall and rainfall according to WMO guidelines. The water levels that are present downstream and upstream of the reservoir must be calculated using an acoustic Doppler current profiler. This equipment can be used to estimate the discharge and velocity. We can also use pressure transducers and automatic millimeters to know the water levels. We need to measure and estimate the velocity of the river at the floor points. This is crucial to know the exact flow discharge. We need to calculate the charge at inflow points. The spillage from the spillways must be estimated along with seepages and evaporation. These parameters can be measured indirectly. We can use a rating curve for this and it is based on water levels that are being observed at discharge levels that are being recorded. We need to consider hydraulic, hydrological and meteorological modeling software. We can estimate the soil loss along with bank and slope erosion. While calculating these parameters we need to consider dip landslides and shallow landslides that might occur in the upstream segments of the rivers that field reservoirs.

Some of the direct methods that can be used to estimate these parameters are reconnaissance methods, volumetric methods and bank shift methods.

In the case of reconnaissance methods, we can measure the changes in the soil surface. This method is simpler and cheaper. In the case of volumetric methods, we can measure gullies and rills along with cross-sections and lengths. The bank shift method can be used with the aid of satellite images and ground measurements. We need to measure deposits at reservoirs, sediment traps and check dams to know the surface erosions. This can be considered as an indirect method. We can also perform image analysis using Remote Sensing imageries. It is essential to model catchment erosion using the revised Universal soil loss equation.

We can also go for direct Sampling and pumping. We can use delft bottles and bags in case of direct sampling. Turbidity meters that are fixed and portable can be used for this purpose. Using several instruments, we can compute the sediment discharge accurately. We can also perform laboratory tests to know the actual concentration of the samples. Some of the laboratory methods are the settling tube method, evaporation method and filtration method. All these methods can be classified as direct sampling methods. We need to know the size of the grain and concentration of the grain in the water segment mixture. It is always important to compute the sediment discharge accurately. We need to understand the suspended sediment using rating curves. This rating can be dependent on the sediment concentration and flow discharge measurements that were recorded using in situ instruments. We can use optical or back scattering instruments in order to measure the grain sample size at the water sediment mixtures. These optical techniques can be useful if the size of the particles is in between 200 micrometres to 400 micrometres at a concentration of about hundredgrams per liter. The optical tools can also be employed in transmission modes. This will be based on the scattering of light and absorption of light by the particles that are present in the sample. The values obtained then give us accurate values that can be used for calibration purposes. We can use empirical calibration techniques here. The nuclear measurer is another device that can be fixed and portable. It is purely based on the back scattering of radiation and attenuation of radiation. We can use this device if the concentration of the particles is in between 0.5 to 12 grams per liter. You can also use acoustic sampling methods to know the properties of the sample given. This sampling method is based on the backscattering properties of a portion of the sample. This technique basically uses sound properties exhibited by the particles.

Laser diffraction is another method that can be widely used in knowing the concentration and properties of the samples. We can use pressure differential also and it is based on the sensitivity of the temperature gradients, turbulent effect and transducers. We can also use time domain reflectometry for this observation. The optical backscattering method, the optical transmission method, nuclear measurer, acrostic sampling, laser diffraction, pressure differential method, digital imaging and time deflection reflectometry can be considered as indirect methods. We can also use our acoustic monitoring techniques such as hydrophones, geophones, dual-frequency acoustic and acoustic Doppler methods. Weight sensors can be used to monitor scour at bridges and the migration of beds. In some areas, pressure difference type samplers are being used. We can also use portable bed load traps and permanent bedload traps. Tracer methods and nuclear gauges can be used for this purpose. The tracer method, nuclear gauges, permanent bedload traps, portable bedload traps, pressure different types of samplers, weight sensors and acoustic monitoring techniques can be considered as Direct methods. We can analyze the temporal variations of bed levels using echo sounders and bathymetric surveys at high resolutions.

## 6.6 SEDIMENT TRANSPORTATION

We need to measure the total sediment transportation that enters the reservoir. This can be done with indirect sampling tools. We need to know the suspended sediment, sampling methods, concentration analysis, size analysis of grains, hydraulic parameters and temperature measurements using computational tools. We can use a settling tube method if the concentration is less than 10000ppm. The diameter of grains should be less than 1 millimeter. We can use the pipette method if the diameter is less than 0.0625mm. In this method, concentration should be less than 10000ppm. We can use a hydrometer for a densimeter if the grain diameter is less than 0.0625mm and the concentration is less than 160000ppm. There is another method that can be used in the analyses of suspended sediments according to the grain size (Martin., 2013). We can use a Visual accumulation tube, dual-frequency acoustic experiment, laser diffraction particle size analysis, Photon correlation spectroscopy and laser diffraction. The sediment characteristics are very important to understand the decomposition of the sediments and their nature. In order to estimate the sediment characteristics, we need to depend on the grain size of the bed samples. We can use the air jet sieving method, wet sieving method, densimeter, pipetting, visual accumulation tube, settling tube method and image analysis methods. The bedstrata of the reservoirs can be studied using side scan Sonar, sub bottom profiler and core samplers. We can use the gravity Core sample also. For the deposited sediment, we can calculate the bulk density. This source of sediments must be determined.

This process involves the collection of soil samples and sediments that are present in catchments and reservoirs. We can also collect sediments and soil samples from the tributaries and river reaches. All the samples can be used to know the mineralogical features and lithology features of a specific site. This can be done using x-ray diffraction studies. We need to analyze the mineral content of the samples procured from the study area. It is essential to know the morphology of reservoirs and rivers. We must understand the bed topography, reservoir bed special features, and deep channels that have longitudinal profiles such as bed forms and sediment delta. We need to understand and evaluate the bed levels at critical zones and that means the intake points. We can use a total station and 3D laser scanner in case of dry zones. You can also use a fish finder in this case. Echo sounders can be used to know the morphology of the reservoir and river. In the case of echo sounders, we can use lead line, single-beam, multibeam interferometric configuration. We can also use image Technology, Radar and LIDAR Technologies. We can use laser profilers and devices that are installed on unmanned aerial vehicles. The quality of water and sediments in the reservoir must be studied. Physical properties should be studied of water quality and sediments letter being deposited. Physical properties constitute taste, smell, turbidity, temperature and color. Along with physical properties we need to consider bacteriological properties. Some of the chemical properties that are to be considered are biological oxygen demand andpH of water that is present in the reservoirs. We also need to understand and estimate the quality of specific sediments. We can use different sensors to know the physical and chemical properties of water. Some of the sensors can provide us real-time measurements and some can provide periodical measurements.

It is mandatory to assess the hydrological process and Hydraulic processes that exist in the River system and catchment that are associated with cascades of reservoirs. Studies and observations of reservoirs and dams must be made available to engineers and researchers as and when required so as to modulate design. Some of the hydrological processes that are linked with catchments are surface runoff and rainfall. These two parameters can be used to measure the inflow rate of the reservoir and erosion rates. There are several automatic weather stations that can be used to know the precipitation patterns and probable runoff magnitudes. We can use this data supplied by automatic weather stations to compute and arrive at meaningful conclusions through numerical modeling.

If the number of weather stations is unlimited then we can use Global Meteorological data provided with the Agencies that are valid. Most of the Global metrological data will be extracted from the satellite imagery. The weather data will be produced using numerical weather prediction algorithms. These models can provide coarse resolution and it is our duty to downscale the results so as to localize data. We can compute the data provided by weather stations and satellite imagery to produce models based on numerical weather patterns. We can also calculate runoff using analytical approaches and other hydrological models. We can compute runoff in relation to the inflow of the reservoirs. The main condition here is that all the catchments must be located at the mouth of the reservoir. The processes about rivers and channel hydraulics are very important. It is essential to have hydraulic data that is linked with water levels and discharge levels at the downstream portion of the reservoirs. Most of the models can give us an exact and accurate prediction of the Reservoir inflow and outflow periodically but it will be cross-referenced with in situ measurements. The observation data is very much essential in the modeling process. The observation data constitutes velocity, discharge and water levels. We need to make impact assessment studies and feasibility studies using these models and what are the data which is available from the site stations or the ground stations. Most of the handling of these processes varies with the regions.

In some regions, flash floods can change the estimations brought about by the modeling process and this is common in tropical countries. The transportation processes and sediment sources may vary based on the hydrological phenomena and the characteristics of catchments. Some of the characteristics are based on river bank erosion, debris flow, mudflow, landslides, gully erosion, slope erosion and the erosion of Agricultural landscapes. Catchment erosion is another important phenomenon that needs to be studied whilst understanding river morphology. The inflow of rivers into the reservoir can be based on the runoff of the catchments. Most of the erosion that occurs in the catchment zones then brings about sediment loads to the reservoir. There will be bedload transports, wash loads and suspended loads. There are several types of directions that bring about sediment loads into the reservoir. Glacial erosion generally occurs in few places where mountainous regions are dominant. Most of the debris from these mountainous regions can be transported to the reservoirs through flash floods caused due to melting of snow. We call such a type of flash flood in the Glacier dominant landscapes as glacial lake outburst floods. Sheet erosion is common in most tropical countries. In case of sheet erosion, the soil particles will interact with the Rain drops and get transferred down the slope or along the gradient. They may not follow the channels that are previously defined or rolls instead. Such a type of reaction can be called sheet erosion. Most of the top soils that support agriculture will be removed due to sheet erosion and so the agricultural fields may lose fertility. Non-cultivated lands will also suffer from sheet erosion. Rill erosion can take place in local zones further classified as small washes. This erosion can be detected through different channels. The depth of these channels may not be greater than 25 centimetres. Because of the downward movement of water along the gradient there will be the formation of gullies. Hence this type of erosion can be called gully erosion. Landslide erosion can damage most of the landscapes that zones and also tropical regions where rainfall is abundant and mountainous regions are available. Which erosion can occur due to the failure of slope naturally or due to Anthropology project activities that involve deforestation and construction of roads? The landslide Dam outburst flood and landslide erosion material may result in formation of larger sediments that might be transported through rivers. Wind erosion must be considered in this case. The surface of the soil is well exposed to the means. Interest of strong winds there will be completely denudation soils. catchment erosion may occur due to several factors namely forest fires, catchment areas, land use, excess soil erosion, topography, geology, lithology and other climatic conditions of a specific area.

The transport of sediments to the reservoirs can be dependent on the river discharge. Trapping efficiency and temperature are to be considered to know the sediment transport in the downstream reservoir. Mostly the temperature and trapping efficiency of the reservoir upstream will be taken into consideration to know the transportation rate of the sediments in the downstream reservoirs. We need to ensure that both this upstream and downstream reservoir share the same river Basin. In many cases, the sediment mass that is being transported can be measured in terms of kilogram per unit catchment area per unit time. This unit can be used to measure the bathymetry of the reservoir and also in quantifying the sedimentation rates. Sedimentation rates can be linked with the erosion rates of a particular catchment. The measurements should be taken in regular intervals and storage components must be taken into consideration to estimate the soil erosion rates at a particular catchment area. We can estimate catchment erosion more accurately using Remote Sensing techniques. The River system picture and disturbance may have balanced the sediment transport and respective morphodynamics processes. It is often difficult to know the source of sediments that may reach a reservoir. It becomes more problematic if there is a mountain in the catchment area. The type of sediment transport can be defined using different factors. We can classify this sediment transport type based on the transport mechanism. We can call it bedload and suspended load. Bed loads can be formed due to jumping, sliding and rolling. The suspended loads can occur due to turbulence and upward flows. They can be classified based on their Origin. They can be divided into watch load and bed material load. Generally, the wash load comes from upstream and it comprises finer particles. The bed material load can be coarser but it is of local origin. You can also classify based on limiting factors. They can be divided into supply Limited transport and capacity Limited transport. Sediment transport adapts to the change in flow conditions in some cases. At that instant, we can classify them into nonequilibrium transport and equilibrium transport. The non-equilibrium transport can be called as retarded adaptation where as equilibrium transport can be called as immediate adaptation.

Most of the larger reservoirs can constitute elements that are coarse and finer grade fractions are not common. Sediments can be deposited in the upstream area and it might contain a coarser fraction of grains. This can be called a Delta feature. Some of the sediments may form thick portions close to the dam. The deposits may be thin in the upstream area. This can be called a wedge-shaped feature. Due to the transport of finer sediments, the wedge-shaped feature may surface. The sediment transportation can be due to turbidity currents in general. This condition is quite common in the case of small-scale reservoirs where sediments are transported easily. If the sediment deposits form as thin layers towards the dam then we can expect the tapering feature. This can be witnessed in longer reservoirs. Rarely we can find uniform features in the reservoir. This can be due to the transport of low fine sediments and frequent fluctuations in the water levels. We can observe combined features in most of the reservoirs.

Chapter 7

# River Pollution

## CONTENTS

7.1 Introduction .......... 116
7.2 Physical Remediation .......... 119
7.3 Chemical Remediation .......... 120
7.4 Dissolved Oxygen .......... 124
7.5 Agriculture .......... 128

## 7.1 INTRODUCTION

River pollution (Figure 7.1) can damage human health and the environment. Most of the regions do not have strict regulations to protect river water quality (Ellen., 2014). It is always essential to maintain environmental integrity to safeguard river ecosystems. Every country should make certain environmentally friendly frameworks so as to save water-based ecosystems. Most of the rivers that flow through the industrial areas exhibit high rates of pollution concentrations and this will wipe out several animal and plant species from the surface of the earth. These pollutants can be detrimental to human health. We need to study the effects of river pollution in detail to know the future problems that are intertwined with the improper management of river ecosystems. Most of the rivers have become inseparable commodities for human societies.

**Figure 7.1** River pollution.

**Source:** *https://en.wikipedia.org/wiki/Water_pollution_in_India#/media/File:Oshiwara_river.JPG*

The riverine ecosystems (Figure 7.2) are being compared with several inorganic pollutants, organic pollutants, dyes and other effluents. Rivers that are loaded with these pollutants will ultimately reach oceans. The Marine ecosystems apart from creating environmental problems and the economies that are based on river-based products will also be affected.

**Figure 7.2** Riverine ecosystems.

**Source:** *https://en.wikipedia.org/wiki/River_ecosystem#/media/File:Stream_in_the_redwoods.jpg*

Most of the fishing communities and aquaculture-based businesses will be affected because of high pollution loads in the river systems. This will also create food problems within the coastal communities. The Exports of the food products or foods that are based on river ecosystems will be drastically reduced and hence the country's GDP will be decreased. There are different types of river pollution. River pollution mainly occurs due to gross solids and trash. The materials that pollute rivers in this case are materials generated from plastic packaging, metal containers and Styrofoam. They can persist in the environment for many years. They can clog several waterways. They can also kill fish and birds. When birds and fishes swallow this material, they will not be able to respire properly and hence they die. Fine solids and sediments can also damage river environments. Grit and sand that is brought about by strong water can get accumulated in the river systems. Because of siltation, there will be clogging of rivers as the total suspended solids increase. Because of the increase in deposits the fish eggs and invertebrates will perish. Determination of chemicals in the rivers can be explained in the form of eutrophication. Most of the chemical pollutants can damage river ecosystems as they come from different sources. These Chemicals arrived at the river environment with varied concentrations along with different materials brought by stormwater or surface runoff. Most of these chemicals will be generated because of the industrial activities at that river basin. These chemicals can harm the river ecosystems and they even get accumulated in several life cycles of animals and plants. Because of their accumulation in plants and animals, humans also will get affected when ingesting them in the form of food materials. There will be in balance in the nutrients and hence in the ecosystems and this will cause algal blooms (Figure 7.3) and which might be toxic to the local aquatic life.

Some of the chemical pollutants are ammonia, fertilizers, detergents and surfactants, hydrocarbons and oils, organic solvents and heavy metals. Fertilizers that are made up of phosphates and nitrates can be found in the river systems that are heavily polluted. We can also observe heavy metals like arsenic, lead, mercury and cadmium in the rivers. Some of the biological contaminants can be seen in the rivers. Some of the pathogens can get populated in the river environment and cause several diseases to humans. The Nigeria river delta can be taken as a good example to represent highly polluted rivers. The Nigeria river is always occupied with shipping transport vessels from and to Nigeria. The oils that are being transported through this Niger are a big problem as spillages and accidents have caused a lot of problems to the local communities in 1958. The Ganges in India is another typical example that explains river pollution. This river is of spiritual importance to the local public and it was stated as the most polluted river in the world. Mostly this river gets polluted because of open sewers, rituals, etc. Industrial wastes and fecal contamination have damaged this river and it has become a hazard to human health. The Citrus river in Indonesia can be classified as one of the most polluted rivers. It is about 300 kilometers in length. There are about 2000 industries installed near the river. These industries have released increased levels of arsenic, mercury and lead into the river. Rio la Matanzas of Argentina is another prominent river that can be considered while exploring river pollution. Most of the river was polluted due to industrial facilities and factories.

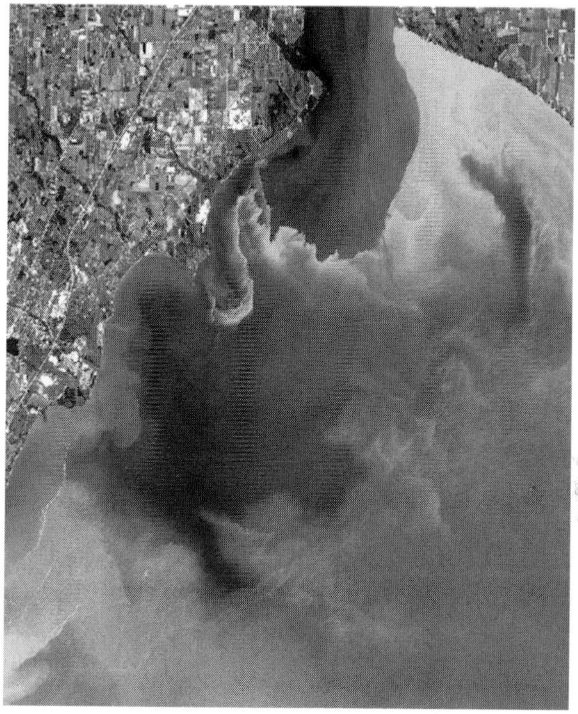

**Figure 7.3** Algal blooms.

**Source:** *https://wcwc.ca/cyanobacteria-cyanotoxins/*

A river can be defined as a natural stream of water that can get emptied into lakes or oceans. There will be the convergence of tributaries all along the course of the river. Streams and rivers can act as Natural drainage systems that can transfer water supplied from the upland areas. Most of the moving water can carry decomposed pollutants in diluted forms. Standing water may contain concentrated pollutants. The domestic, agricultural and industrial wastes are being dumped into the rivers. Due to the rapid development of the economy, most of the river ecosystems are being damaged seriously and this is much closer to the tipping point where we cannot rejuvenate the rivers in the near possible times.

## 7.2 PHYSICAL REMEDIATION

Physical remediation is an important process in the rejuvenation of rivers that are being polluted. Aeration is an important technique that can be classified as one of the physical remediation techniques. It can restore the growth of microbes that improve the water quality in all the rivers. This can be classified as an effective and simple technique. While using this technique we need to consider the characteristics of the source and the channel conditions. The water improvement needs to be taken into consideration while using this technique. There are several techniques in the aeration process namely water aeration, mobile radiation and fixed aeration (Frankel., 1978). River aeration technology has become popular in most advanced countries because of several practical applications that can improve water quality artificially. This technique has been proved in the treatment of the Oeiras river of Portugal. We can control river pollution effectively using a water diversion method. This method was used in major cities of China. This technique includes the diversion of effluents before reaching the river systems. It demands higher costs but the pollutants that are being flushed out will be large. Sediment dredging is another important method that can be used in the treatment of river water. There are several dredging methods that can bring about determinantal environmental effects. We can use a drag suction dredger or grab suction dredger in most of the operations in the river ecosystems and this may cause sediment resuspension. A water injection dredging method can be used if sediment dredging is to be done on the site. This can effectively improve the efficiency of the process and can reduce costs. This methodology was used in the UK. The cutter suction dredger was considered an advanced type of environmental dredging method. It has several monitoring systems that can increase the accuracy of dredging.

## 7.3 CHEMICAL REMEDIATION

Chemical remediation is another technique that can be used to purify the river waters. In this technique, we can use two methods called flocculation and sedimentation. These methods can be used to treat water that is already populated with suspended solids, live algae and others. Which chemical remediation process is easier to operate (Hossain et al., 2020)? It can be easily used to treat river waters that are already in a secondary pollution state. It can be linked with the pre-treatment processes so as to decrease the load at the river mouths. The algae can be removed by chemical agents. We can use copper sulfate to remove algae present in the river environment. Some of the other chemicals that can be used to remove the algae s are ferrous sulfate, poly aluminum, alum, bleaching powder, etc. we can reduce the smell of water using these chemicals as the algae will be eliminated and as their products produced by the algae will also be removed. The pH of the water will be changed with the addition of copper sulfate in the river systems. We can use calcium hypochlorite to kill the green algae, diatoms and cyanobacteria which can cause water blooms and create toxicity in the river environment. Bioremediation techniques are also very important as far as the river system pollution control is concerned. In the bioremediation process we can use cultivated microbes are indigenous microbes and several other related organisms to transform the toxic compounds into nontoxic compounds and harmful pollutants to manageable pollutants and so the local environment can be protected. The variation process was first attempted in Pennsylvania to minimize the gasoline pipeline leak in 1972. It was considered as the first-time application of this bioremediation process at a large scale and it had brought about much attention during this.

The bioremediation process can be classified into natural bioremediation and artificial bioremediation. The artificial induction process can be classified into Ex-situ bioremediation and in situ bioremediation. Mostly in-situ bioremediation can be used to treat polluted rivers without transportation and excavation of the pollutants. This is very much useful as far as the pollution control of river systems is concerned. We can use indigenous microbes along with domesticated microbes to reduce the pollution load of the rivers. There are several methods that are associated with this process. Ex-situ bioremediation process may involve the transportation of contaminants from the river site to the treatment plant and this is a hard process. There are advantages to this process. It is cheaper and can cause minimal environmental problems. There will not be any pollutant transport for the formation of secondary pollution. Most other pollutants can be neutralized and hence when you are mentally protected at that site only and the rivers downstream. This can be used as a regular pollution treatment technique. We can also remediate the polluted water using remediation Technology that can be assisted by aquatic plants. Some of the plants can purify water pollution in the rivers and they are called aquatic plants. They can observe pollutants and can tolerate higher concentrations of polluted river waters. The water pollutants can be eliminated through the degradation process of plants, accumulation, absorption and adsorption processes. We can use water hyacinth and other plants to remediate the polluted rivers.

There are many simulation experiments and laboratory experiments that are being done in several Laboratories pertaining to river Pollution Control. Aquatic plants can effectively remove nitrate, nitrogen, total nitrogen and total phosphates that are being deposited in the rivers. We can use aquatic animals to remediate rivers that are already eutrophicated. In this case, the identity and composition of the features will be altered so as to bring about a healthy river environment. We can use filter-feeding fish, common carp and silver carp to remediate eutrophicated rivers that are populated with phytoplankton.

We can use microbial fermentation technology to reduce the pollution load in the river environment. Biofilm technology is one of the popular technologies that can be used in many places. In this technology, we can use bio membranes that are linked with the natural river bed. A micro carrier will be installed so as to transport the pollutants from the rivers. Biofilms use filtration, degradation and adsorption processes to reduce the pollution load in levers. The effectiveness of the biofilm technique will be based on several external conditions and substrate compositions. The biofilm technology was successfully used in South Korea and Japan. Some of the techniques that are based on biofilm technology are underground stream purification technique, thin layer flow method, artificial packing contact oxidation method and gravel contact oxidation method.

Many ceramics can be used to treat pollution water as it can remove algal members, Mn, color, turbidity, etc. Microbe Dosing is microorganism dosing that can be used to transform, absorb and decompose the pollutants that are present in the river waters. The growth of useful bacteria and several other microbes can be considered a challenging task in the case of microorganism dosing.

The causes of Water pollution can be classified into 2 main categories. There are point sources and nonpoint sources. The harmful substances are the effluents from the industries that can be directly fed into the river systems in case of the point source. The pollutants can be indirectly injected into the river environment through other sources in case of non-point source mode of water pollution. The pipes that are installed in industrial facilities can dump their plants into the river systems and they can be classified under point source of water pollution. In some cases, the pesticides and fertilizers from the agricultural fields will be transported through water in the form of surface runoff to reach the river ecosystem. This can be called a non-point source of pollution. we can easily monitor and regulate the point source pollution. There will be a lot of political factors in place if you consider a point source of pollution and its regulation. We cannot monitor and regulate the nonpoint source pollution. Most of the contaminants in the lakes and streams are due to nonpoint pollution sources. The fertilizers and pesticides can damage the liver environment to a higher extent. The nitrates and phosphates can be classified as major polluters of the rivers. Additives, gasoline and oil can pollute river waters.

Mining is another important factor that can pollute river water with heavy metals and sulfur compounds generated due to the excavation of Earth. The rainwater that falls on the surface of these mines will transport acid mine drainage into the river ecosystem and hence results in heavy metal pollution. Rainwater can carry mining waste that tailors freshwater systems. In the mining of gold, cyanide will be used on the mined rock to extract gold. This cyanide will be transported to the nearby water systems. Most of the mining waste will be stored in the containment dams in the form of slurry. Industrial and chemical processes can also contribute greatly to water pollution. Water pollution can be caused due to sewage disposal. Human infections are common if water gets polluted with pathogens. There will be the generation of water-borne diseases due to pathogenic populations that may develop in the contaminated waters. Most of these pathogens can be connected to river systems through sewage disposal and excreta from humans. Water-based diseases can also create a lot of problems for human beings.

The photodegradation of water pollutants can be useful to save most of the water bodies. Most lead pollution can be associated with several anthropogenic activities that are linked to generate goods and services for the population. River pollution can be due to harmful chemicals that can be organic and inorganic along with microbial infections and populations. The advanced oxidation process was one of the popular methods used to treat the wastewater. In this process hydrogen peroxide is used. Ultraviolet radiation and ozonation are also used in this process. Ozone exhibits higher oxidation potential and it can react with its components slowly. It can react with chlorinated herbicides and chlorinated alkanes. This sophisticated treatment method can combine the use of ultraviolet light and ozone. During this process there will be the production of hydroxyl radicals. These hydroxyl radicals are very reactive. We can use Ozone alone to treat the wastewater that is being generated from the paper and pulp industry. We can also combine biotreatment with the ozonation process. To remove trichloroethylene and choro alkanes we can use hydrogen peroxide and ozone successfully. The treatment process that uses ozone gas can be expensive.

## 7.4 DISSOLVED OXYGEN

Dissolved oxygen (Figure 7.4) present in water can work as the best oxidant to degrade pollutants that are present in water (Liptak et al., 2018). We can use photocatalysts such as Titanium dioxide also. Titanium dioxide-dependent photocatalysts can be used repeatedly. Sunlight can naturally degrade the pollutants that are present in river environments. Most photocatalytic reactions can be classified as exoergic processes. There will be the acceleration of electrons due to the use of energy from the sunlight. Most of the organic pollutants are resistant to photocatalytic reactions. There will be delayed oxidative degradation. The photochemical process regulates many reactions that might occur in the upper atmosphere. Most photochemical reactions are being bypassed by photobiological and biological reactions.

The molecular oxygen can be classified as a poor oxidant. This is the reason why we need to opt for a photo assisted oxidation process, especially in the aqueous media. Most of the river environment can contain reactive transient species that can react with the photochemical reactions. Flash photolysis techniques can be used in river ecosystems. The primary intermediates in the photo oxidation process are hydrated electrons, hydrogen radicals and singlet oxygen. We can also get superoxide radical anions that are a very much crucial decontamination process. Trapping of hydroxyl radicals can be achieved by the spin trapping method. Organic contaminants that are present in the river systems can be decontaminated using a photolysis process. This is especially useful if the organic contaminants are in low concentrations.

There are several routes for the pharmaceutical molecules to reach the rivers. These can reach via point source and nonpoint sources. They get deposited on sludge and wastewaters. Most of the wastewater treatment systems use an activated sludge process because effluents produced in this process are not harmful. operation and maintenance cost for much lesser. In the preliminary treatment, we can use a grease removal unit, oil removal unit and grit chambers along with bar screens. There will be a primary gravity settling tank to purify the effluent. The wastewater that is dominated by pharmaceutical molecules may be subjected to a biological treatment process. This process is facilitated by aerobic bacteria that are being suspended in the form of beds.

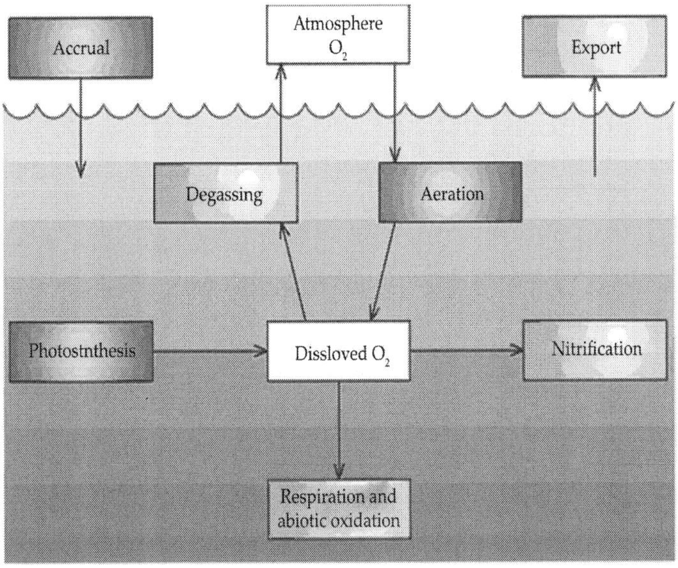


**Figure 7.4** Dissolved oxygen.

**Source:** *https://ozcoasts.org.au/indicators/biophysical-indicators/dissolved_oxygen/*

All the rivers irrespective of their type mainly follow the same stages of structural variation from the source to the destination. In general, rivers can be divided into three primary zones. They are the source zone, transfer zone and deposition zone. This source zone can be called headquarters. It constitutes mountain streams that exhibit Rapid slope patterns downhill and there is a formation of V-shaped channels. In some cases, waterfalls may be produced. There will be the transport of large sediments downstream the rivers. The transfer zone can be represented by low altitude. At this zone, the velocity will be less. The river bed will start the meandering process and become wider. There will be the settlement of large sediments at the transition zone created at the junction of the source zone and transfer zone. There will be the production of sediment cones in this area. Some of the sediments can get a transporter downstream the rivers. The deposition and erosion process will be balanced at the transfer zone. The deposition zone can be represented by having a less slope. The flow velocities will be reduced at the deposition zone. There will be a meandering process in this zone. Fine sediments also get settled in this zone. The mouth of the river can get opened in the form of a delta where sediments are dominant at the bottom of the delta. At this point, the river gets divided into two arms. The natural rivers may not be straight. We can use the sinuosity index to know the shape of the river. This index can be calculated by dividing the middle river length by the actual length of the valley. If this index reaches 1.3 then we can assume that the river is subjected to meandering.

This index may vary with the different sections of the river. If the shape of the river is changed due to anthropogenic activities then the meandering process will be halted. The river will try to restore its original shape due to sinuosity. Pools and Riffles can be considered as essential elements and the river systems. Almost all rivers exhibit sequencing of deeper sections and shoals. The sequencing is equidistant. The pools and riffles can be associated with the midstream of the rivers. The midstream is nothing but the deepest portion of the river. At the midstream, the flow velocity will be higher. Pools can be generated at the mid-stream of the river. This can be outside the bank of the natural bend of water course. The difference can be produced between two naturally occurring bends. Generally, the midstream can proceed from one riverbank to the other river bank. The distance between any two riffles or pools must be calculated to estimate the width of the river. It will be at least 5 times the width of the river. The Sequencing of these entities can help us understand the biotic communities that can survive in both slow currents and fast currents in a river ecosystem.

We need to understand the horizontal structure of the river corridor. The is the important part of the river corridor riverbed. This is nothing but the passage for flowing water. Floodplains can exist very near river beds. In some cases, the flood plains may be present on both sides of the river bank and it is based on the relief patterns. The upland fringe can be considered as an important component and it starts as both the floodplain. It can form a frontier between adjacent landscapes and river corridors. The bottom portion of the river can be considered as a river bed. The size and shape of the river bed can be studied based on the balance between the four basic factors. We need to consider the energy of the water stream based on water velocity and slope. The substrate generally offers resistance against the flow. We need to consider runoff and the size of the sediments. The particle size of the sediments must be calculated. Different plants can be classified into topographic floodplains and hydrological floodplains. The hydrological floodplains are nothing but an area that is being inundated frequently at least once in 2 years. Topographic floodplains can constitute hydrological floodplains until a certain altitude is reached. We can use topographic floodplains in the planning and development of river systems. This planning is important in forecasting flood management practices.

Most of the floodplains that are present in the downstream areas can be considered as dynamic systems. There are several components that are to be mentioned that explain the formation of flood Plains and their disappearance. Meander Scrolls regain important features that can be seen in the river ecosystems. In the meander Scrolls, we can observe alluvial deposits. These deposits can help us understand the properties of previous riverbeds. The abandoned river bends or oxbow lakes every form due to gradual detachment from the main river system. That shape will vary according to the development stage. Some structures mainly form at high waters if the streams pass through natural levees. Some of the spread deposits can exhibit fan-shaped structures. It can be called a splay. There will be the formation of wetlands after a high-water situation due to the formation of natural levees. This can avoid water or stop water from returning into the river bed. This can be called a back swamp. Due to this, there will be an extended flooding period. Most of the wildlife in the river systems will be based on the components of the flood plains. Most carp species require oxbow lakes and back swamps for their reproduction and growth. It is to be mentioned that damage to riparian wetlands reduces the fish population downstream. The vertical structure of the river is also important at the horizontal structure of the river. In general, the level of water will vary according to the season.

Variation will be anywhere between 10 to 20 centimeters in normal conditions. The increase in the water levels can be attributed to the seasonal variations and it will affect the species that grow in the river ecosystems. Most of the rivers are linked with the adjacent wetlands when the water levels are increased. This situation is much needed for the fish population so that they can migrate to other places easily. Some of the specialized vegetation systems may develop at the river banks. This can be attributed to the water level variability at the banks of the river. The floodplain forest can be rich in animals and plants. This is seen in equatorial jungles. The depth of the river may not be calculated only to its bottom. The riverbed cannot add water-saturated zones also. Groundwater can floor into the river valleys and from the river valleys. The stream that is present in the river bed can be a small portion of water flow that can run through the valley.

Groundwater levels can be based on the water quantity of the river bed. In seasons where there is scanty rainfall most of the river bed gets water from the surrounding areas and so the groundwater will reach the river bed.

In case of flooding situations, the water will flow from the river bed to the exterior and also in the form of groundwater. Rivers can be stated as migration corridors. Most of the wildlife can migrate for breeding and feeding purposes both downstream and upstream of the river. The migration mechanism will occur due to an avoidance mechanism followed by the species inhabiting the river water. The vegetation that is present in the catchment basins can slow down the river water runoff that may reach the river and so the high tides that may be present in the river can be nullified. Access visitation in the catchment area can also lead to reducing the evaporation from the surface of the earth and hence the groundwater will be maintained properly. Recorders can be considered as natural links between the isolated forest areas and other areas that can support plants and animals indefinitely. The quality of the river must be studied periodically systematically to come up with meaningful conclusions.

## 7.5 AGRICULTURE

Agriculture contributes to many water-related problems. It might cause sedimentation and erosion in the water bodies (Bent., 1968). The pollution that may be caused due to sediments can be classified into physical dimensions and chemical dimensions. If land gets degraded and there is a loss in the top soil resulting in sheet erosion and gullying, there will be increased turbidity in the rivers. This can lead to physical and environmental impacts on the lake beds and rivers. If the clay fraction and cell fraction are higher then they can carry chemicals through the adsorption process. Through this process, the chlorinated pesticides and phosphorus-based pesticides along with metals can be transported in the form of sediments to the aquatic systems in the nearby water systems. The loss of topsoil due to erosion can lead to loss of metal and hence organic matter will be completely replaced with fertilizers-based chemicals. The soil productivity in this case will be based on the fertilizers and pesticides added by the farmers. This will lead to physical, social and economic consequences in the societies, especially the farming communities. It is important to control agricultural pollution by regulating sediment runoff and measures that are being used to control erosion. In some cases, sediments can be treated as physical pollutants per se. Reliable values of sediment concentration and discharges are always questionable as many countries will be reluctant to offer information regarding their agricultural activities. The sediments can affect the waters that are being received by the rivers. There will be an increased level of turbidity and this leads to decreased sunlight penetration into the water column. The decrease in sunlight penetration can increase algal members and other aquatic plants that are rooted in the deposits of river ecosystems. Most of the gravel birds will be covered with fine sediments that can damage the environment needed for the spawning of fish. The aquatic system will be damaged due to habitat loss. The heterotrophic nature of the river will get increased and thereof increase the growth of algae and plants that are rooted. These consequences can increase the turbidity levels of the rivers. Because of increased sedimentation levels, the hydraulic characteristics of the channel will get disturbed. This will impact the navigation properties and hence the channel that will be affected. Because of this there will be enhanced flood rates as the storage capacity of the river channel will be effectively reduced at the drainage basin. There will be blocks in the hydraulic facilities and river transportation will be disrupted.

The sediments can be treated as chemical pollutants. The effect of sediments will be based on their chemical properties. This chemical property can be attributed to the particle size of the particular sediments at the organic carbon content. The silt and clay fractions play a crucial role in determining the chemical activity of the sediments. Metals and phosphorus-based deposits can attract ions exchange with the clay particles that are present in river systems. There will be contaminants that are considered organic due to the bioaccumulation process. The octanol and water partition coefficient can be used to know the affinity of the particulate matter with organic chemicals. Degradation of sediment organic chemicals can be based on microbial degradation patterns that are common in sediment transport and deposited sediments. Most of the organic chemicals can get linked with sediments and reach the food change in diverse ways possible. Segments can be absorbed by the fish and the final segments can reach the higher systems. These can be classified as food sources for fish. Most of the toxic compounds can get accumulated in the fishes resulting in biomagnification. The fundamental controlling factor of surface runoff is the intensity of the rainfall. This intensity of rainfall determines the amount of water the gas deposited on the surface of the earth and it is closely known as the soil detachment rate. The detachment of soil can render the soil particles freely available for sediment runoff. Soil permeability is another factor that is to be considered to determine the rate of runoff. Soil permeability can be considered as a physical character of the soils. The tendency of the soil that gets impregnated into the interior layers of the ground can be considered as permeability of particular soils.

Chapter

8

# Reservoir Pollution

## CONTENTS

8.1 Introduction .......... 132
8.2 Chemical Characteristics .......... 135
8.3 Microbiology .......... 140
8.4 Water Quality .......... 141
8.5 Operation .......... 142

## 8.1 INTRODUCTION

The Reservoirs can be considered important water bodies that are generated by humans for determined purposes. They are constructed to regulate rivers and utilize water for meaningful purposes (Huang., 2009). The important uses of reservoirs are to supply municipal water, drinking water, water for cooling equipment in industries, generation of power, irrigation, flood control, regulation of rivers, recreational fisheries, commercial fisheries, boating, body contact creation, canalization, waste disposal and navigation purposes. Most of the reservoirs are constructed in areas where water is scanty or excess. If you want to control water, we can use it to revise effectively to bring down the harmful effects of water sources. Reserved words can be used to conserve water for future uses. We can use this water for drinking also. If the water inflow is exceeded then we can use a reservoir to control the floods. There are certain anthropogenic activities that can affect both the quality and quantity of the reservoirs. Some of them are fish farming and power generation. Most of these water bodies can be affected by anthropogenic activities. According to the United Nations environmental program, 1991, most of these reservoirs are rated as undervalued. Reservoirs vary in size and sometimes they may be as large as large lakes and ponds. The morphological variation in the reservoirs can be obvious. We can easily demarcate the morphology of reservoirs and lakes. Most of these water bodies can be assumed as reservoirs. The reservoirs can have different attributes as the lakes. Most of the reservoirs must be subjected to quality requirements if to be used for human activities. There will be differences in the operation and design of reservoirs.

Generally, any water body that is to be contained by a dam or embankment and being managed for community requirements can be called a reservoir. Natural waters can be stored to satisfy the demands of humans and can also be classified as reservoirs. These reservoirs that are being formed with a dam constructed across a river with certain inundations in the upstream land can be called as impoundments. Most of the water bodies can be constructed along the course of the river. Some of the water bodies may not be constructed along with the course of the river and so they'll be enclosed with certain banks called waterproof banks. They are called off-river reservoirs. Some of the reservoirs can be produced by dams in a series along the course of rivers to produce a cascade. The off-river reservoirs and impoundments can be considered as artificially created water bodies. The action of humans on these water bodies will vary e according to the location and morphology. Most of these impoundments can be larger and dendritic.

The primary function of these water bodies is as impoundments to regulate the volume of water that is to be contained. In some cases, the inflow and outflow will be regulated. The off-river reservoirs can be isolated from the local landscape by waterproof embankments. Most of the impoundments can get interacted with floodplains.

We can observe a lot of variation in the Reservoir construction process. The main use of reservoir will vary with time. Some of the other functions will evolve because of regional geological variations. The impoundments can be produced from different Dam types. The impoundments can be classified as buttress, arch, gravity and Earth fill. Earth fill dams can be classified as commonly available dams as 85% of dams constructed come into this category. In case of extremely high wall requirements, we need to opt for arched dams. The off-river reservoirs can be constructed as a whole structure. The Embankments for the structure can be made up of clay and concrete. We can use Earth embankments made up of stone and concrete on the waterside. Most of the reservoirs are constructed primarily to store water. in some cases that can be constructed to transfer inter basin waters. It is always important to know the effect of artificial and natural influences on the reservoirs. This will help us understand the biological behavior, physical and chemical behavior of the reservoirs under investigation. The improvements that are being formed by the dams can be prevalent in the case of large reservoirs. The important impoundments can occur as a series or a single then it is based on the local topography. Most of the impoundments can be dendritic. These cascades can hold water with high quality and hence the water that is present downstream can be used for several human activities as the contamination load is less. We can call the upstream impoundments free impoundments. The off-river storage can be done with the help of reservoirs away from the initial course of rivers. We can operate these systems majorly in series or as single entities. This type of reservoir can be used for water supply. These storage reservoirs can be used to pump drinking water to the household. interbasin transfer of the water resources can be important in reservoir engineering. Inter-basin transfer schemes can be classified as hydrological systems that are artificial and they come with several potential impacts during their operation. Most of these interbasin transfer schemes will be witnessed in regions where water scarcity is high. In this process, water will be extracted from the nearby watershed and it'll be allowed to pass to the nearby watersheds. This transfer can be done with several methods like pumping and networking of canals. In some cases, the intervention transfers can be facilitated by gravity flows. The structure and operation of the reservoirs can affect the water quality parameters and hence these parameters must be adjusted as and when required. There is a system that is constructed in densely populated areas that can exhibit high levels of management intervention than less populated areas. The reservoirs exhibit biological characteristics, chemical and hydrodynamic properties also.

The reservoirs that are produced due to the impoundment of rivers suffer from water quality. There will be the establishment of the latest ecological balance in the reservoirs. Some of the water quality characteristics that are to be considered in the reservoirs are flow velocity, residence time, the mean depth of water, oxygen input, variable oxygen concentration during the day and the night, self-purification of contaminants, oxidation of ammonia, the deposit of fine sediments, resuspension of fine sediments, turbidity, depletion of oxygen mainly in the sediments, secondary pollution and aquatic plant life. We need to consider the thermal characteristics of the reservoir. Most of the reservoirs will have thermal properties like lakes. The reservoirs that are constructed in temperate areas might be deeper than the tropical and polar regions. This can be due to annual cycles that are variable according to the region. There are several reservoirs that have Shallow depths due to enhanced turbulence from the sediments that were deposited upstream of the rivers. Thermal stratification can be based on the local basin geography and other external factors such as climate and weather situations. The reservoirs that were constructed on the floodplains can exhibit minimal depth conditions. The ravines, river channels and canyons can permit the mixing of waters. Due to thermal stratification, there will be a large variation in the water quality parameters. some of the mixings can be attributed to internal waves. These internal waves can be called seiches. These waves primarily occur at the source of the reservoirs. The internal waves can represent epilimnion water. It constitutes lodge algal density with oxygenated and warm water. These internal waves can also exhibit hypolimnetic water. This water can be categorized by the presence of hydrogen sulfide, Mn, ferric ions and the water will be anoxic and cold. Due to these differences, the quality of water will be changing continuously.

Due to increased ambient temperatures in the tropical reservoirs, the temperature range will be minimized and hence the stratification process is not witnessed. Most of the reservoirs can exhibit dynamic inter-annual variabilities than the naturally forming lakes in the local regions. Most of these variabilities can be attributed to water utilization and operational guidelines that can control operational protocols. The impoundments will always represent thermal patterns at different times of the year. This can be attributed to fluctuating water levels that resulted due to diverse operational protocols. The depth can be satisfactory to initiate the stratification process and even Shallow waters can also allow constant mixing at the deep stage levels. The variation of depth can be a standard behavior that is to be considered while understanding thermal stratification. In case of loss of thermal stratification in the reservoirs that are much deeper this segment will proceed to abnormal conditions. Abnormal conditions that are experienced by sediments can be oxygenation and temperature varying at different levels. They can also exhibit stagnation periods in the summer season.

There are certain control facilities installed in the reservoirs to monitor stratification processes. This regulation facility can help reservoirs to maintain the natural turbulence. There will be a modification of horizontal distribution and vertical distribution of the chemical entities. The distribution variance can also be applied to biological organisms. The entire production dynamics of the biological systems will be severely affected.

## 8.2 CHEMICAL CHARACTERISTICS

The chemical characteristics of the reservoirs must be considered as there are several chemical events that occur in the reservoir without our notice. The intensity and timing of the chemical events may be difficult to quantify. The outflow and inflow velocities and the morphometry can drastically influence the water-related characteristics of the reservoir. There will be reduced vertical transport in the reservoir. This suppression can be seen in thermocline regions. At this region oxygen gradient takes place. This zone will eventually form as hypolimnion. The Mn and iron molecules will be supplied back into the water columns and therefore there will be the formation of hydrogen sulfide in large amounts. In the subtropical and tropical areas, we can observe the deoxygenation of hypolimnion. This can be attributed to enhanced decomposition rates that might occur in presence of ambient temperatures. The deoxygenation process can be considered as a burning problem in tropical impoundments. This can be due to imposed oxygen demand needed for the decomposition of carbonaceous materials. The oxygen supply will be limited and hence there will not be stabilized aquatic zones. It is always necessary to take into consideration the substantial draw-offs as they influence the core properties of the reservoirs. The chemical and biological profiles will be severely affected because of this activity (Hai Wu et al., 2012).

The denitrification process can lower the nitrate to nitrogen concentrations. This can bring about instability in the water columns. There will be increased internal turbulence in the reservoir due to the denitrification process. The lowered dissolved oxygen, prolonged retention time and relative shallowness can also affect the turbulence ratios. The biological characteristics of the reservoirs must be studied to maintain turbidity and other properties that can affect the water quality of the reservoirs. Mostly, the biological properties of the reservoirs are the same as the biological properties of lakes. There will be different physico-chemical conditions that might exist in the reservoirs.

Ecological succession and biological production in the lakes can be high whereas in the reservoirs these two processes may occur at low rates. The biological growth rate in the reservoir can be affected by several factors. The turbulence and flow rate can be considered as crucial factors that control the growth of biological organisms in the reservoirs. Nutrient loading is another residential factor that needs to be calculated to know the biological growth rates in the water present in the reservoirs. Planktons can also support the growth of biological species. It is always better to maintain proper management practices in the reservoir systems. The physical and chemical properties of the reservoir pertaining to biological species should always be monitored by different sensors available at the monitoring stations. There are several techniques two assess the growth of biological species. We need to know the interactions between plankton and fishes along with algal communities. Increased turbulence can be due to the destratification methods. These methods can hold plankton in the suspension rather than sinking to the sediment zones where light scarcity is common. In the deep zones, the phytoplankton (Figure 8.1) may not get enough light and hence there will be a reduction in the photosynthetic rates. The enhanced turbulence can produce higher amounts of organic matter that can be available in the suspension.

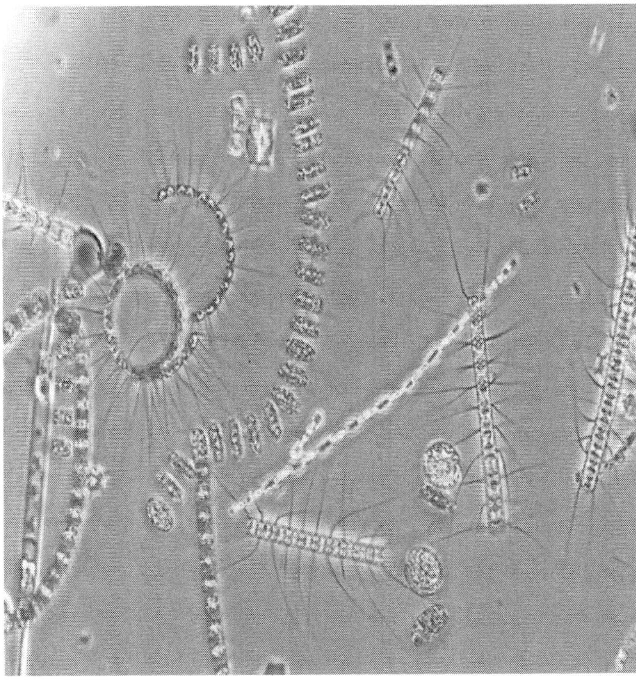

**Figure 8.1** Phytoplankton.

**Source:** *https://en.wikipedia.org/wiki/Phytoplankton#/media/File:Mixed_phytoplankton_community_2.png*

The major problem with the reservoirs is that the driving force that drives the management practices is limited. The effect of biological systems of the reservoir can be dependent on the eutrophication levels. This can be attributed to the increased growth of phytoplankton in the reservoirs. These phytoplankton populations can reduce the recreational value. Most of the consumers will be reluctant to utilize water from the reservoir because of the bad smell and taste due to the growth of algal members. The bundled reservoirs can be used to supply water but there will be the presence of filamentous blue-green algae called Oscillatoria (Figure 8.2).

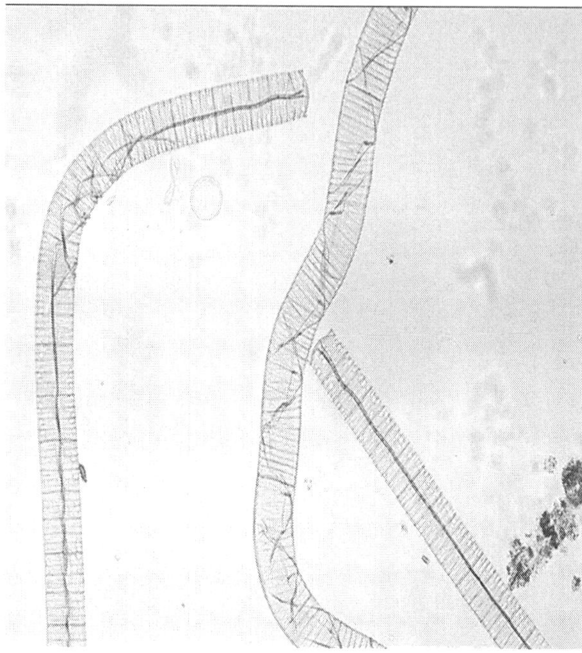

**Figure 8.2** Oscillatoria.

**Source:** *https://upload.wikimedia.org/wikipedia/commons/e/ec/Oscillatoria_princeps.jpeg*

Due to the presence of these algae, there will be a bad odor and earthy taste (Van Breiman et al., 1992). The blue-green algae are cyanobacteria that can generate toxins that are classified as intracellular toxins. These toxins will dominate water if the blue-green algae are decaying and senescent. Mostly, the production of toxins is a natural process and so we need to improve management practices to combat the damage caused by blue-green algae. We can use copper sulfate to control blue-green algae. These toxins can impact both humans and livestock.

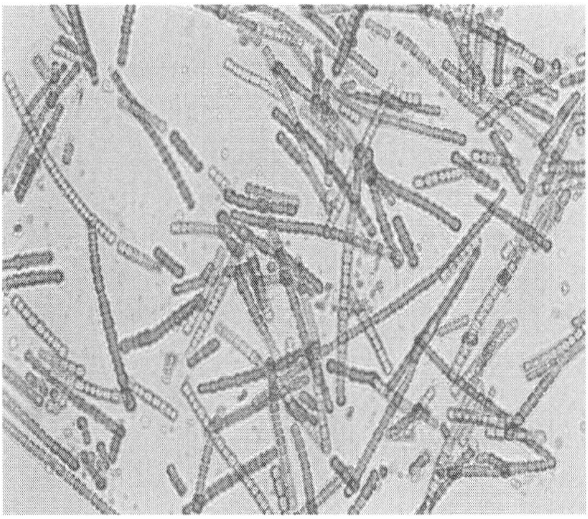

**Figure 8.3** Cyanobacteria.

**Source:** *https://en.wikipedia.org/wiki/File:CSIRO_ScienceImage_4203_A_bluegreen_algae_species_Cylindrospermum_sp_under_magnification.jpg*

The toxins generated by cyanobacteria (Figure 8.3) have become a burning problem in reservoir management. The reservoirs that are present in the subtropical and tropical regions can be classified as warm water reservoirs. The high temperatures of the reservoirs can damage floating plants like Eichhornia and Salvinia. These macrophytes can lower dissolved oxygen in the water. This can be observed while they decompose. Reduction in oxygen levels can lower fish growth and hence this water can be used only for industrial purposes. There are certain standard procedures that can be used in both assessment and management of reservoirs and it will be based on the location and a variety of the reservoir taking into consideration their uses. In the case of natural lakes, there will be trophic relations that exist in the reservoir. There will be food changes that are simple but they represent complex food webs. If a Reservoir is having low-grade morphometry then we can assume that there are no macrophytes and littoral zones. This is the reason why zooplanktons (Figure 8.4) may not grow well. This will lead to reduced supply of food fishes.

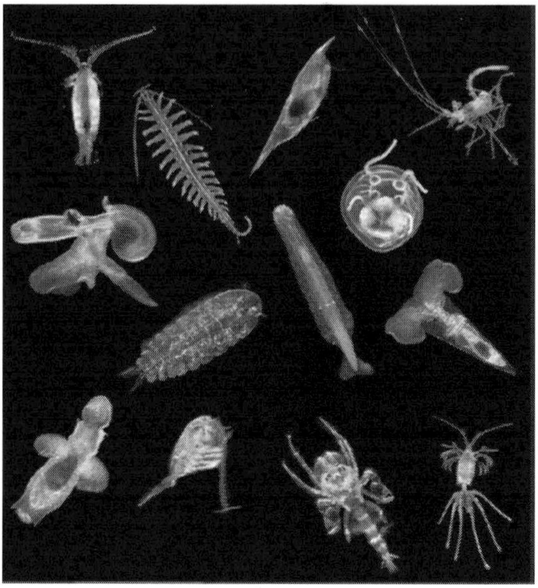

**Figure 8.4** Zooplankton.

**Source:** *https://onlinelibrary.wiley.com/cms/asset/1bc46707-aba1-4e97-affb-d17d83474e36/ece3644-fig-0001-m.jpg*

## 8.3 MICROBIOLOGY

The microbiological properties of the reservoirs are the same as the microbiology of lakes. Bacteria that live in plankton can decompose inorganic material (Hennigan et al., 1996). There will be minimal retention times and hence bacteria arrive into the reservoir from the inflow of water instantaneously. If the retention time is increased then there will be an increase in primary productivity within the reservoir. The bacterial populations can be subjected to regional seasonal variability. The growth rate of bacteria can be related to primary productivity. The activity of microbes can be important in the decoration of several compounds that are being pumped into the reservoir from their industries. There will be a dilution of organic contaminants that can allow the microbial breakdown of the chemicals and this can be important to prepare water to be consumed by humans. This purification process by bacteria main demands sequential operation of the reservoirs. These source waters that feed reservoirs can be loaded with bacterial and viral populations. This can be attributed to surface runoff and effluent disposal. Mostly the reservoirs can act as barriers for most of the pathogenic microbes. The coliforms in the water can be reduced to 1% if the water is stored for more than 20 days in the reservoirs; this can be due to ionic conditions, ultraviolet radiation and the temperature of the reservoirs.

The trophic classification should be studied to know the stratification of reservoirs biologically. Most of the reservoirs can represent diverse trophic States that can be the same as Natural lakes. There will be spatial heterogeneity if you observe the properties of lakes and reservoirs. The topic States Meri presents eutrophic portions and oligotrophic portions. The eutrophic portions can be classified as nutrient-rich portions whereas the oligotrophic portions can be classified as nutrients for portions. Eutrophic portions can be witnessed in the upper portions of the river whereas the oligotrophic portions can be observed in the vicinity of dam walls. These conditions are likely to be enhanced if the reservoir exhibits different morphologies that can be attributed to human settlements and other factors.

## 8.4 WATER QUALITY

Water quality issues can determine the life of a reservoir and its utilities. Several social and environmental issues surround the reservoirs. Eutrophication is considered as one of the more important factors that are to be considered while dealing with their water quality. The reservoirs can suffer from eutrophication like lakes that form naturally. Some of the reservoirs undergo an aging process and hence can be able to convert some of the lakes into wetlands. Thesewetlands can slowly get converted into terrestrial biomes. The design life of the reservoir can be at least 30 years. Most of the devices can be prone to cultural eutrophication because of increased inhibition of human settlements at a given watershed. There will be sterane impoundments that can be classified as protective because of the presence of watersheds at a specific location (Peter et al., 1996). If there are large watersheds then we can expect large sediment loads into the water bodies. The enhanced flushing rate with short residence time can moderate the sediment load transport and this can be observed in the natural lakes. The water load to be deposited into the reservoirs can be based on the watershed area. We need to search for artificial control mechanisms to reduce the eutrophication process in the reservoirs. The chemical-based damage and thermal stratification must be regulated with human intervention. We can reduce primary production by circulating phytoplankton in the water artificially. This can be done under the euphotic zones. The euphotic zones are nothing but areas that can support the production of phytoplankton. Due to this, there will be e reduced photosynthetic rates. Reproduction dynamics of the sedimentation and respiration will be reduced. Phosphorus and nitrogen levels can also influence the efficiency of primary production. Bio manipulation can also help us to regulate primary productivity. The regular control of water levels can also fluctuate the spawning time of the fish population. We can use modified operation regimes that can regulate nutrient loads and water in the reservoir to reduce the nutrient concentrations effectively. We can use scour valves, pre- impoundments and flood bypass to regulate the nutrient concentrations. The river waters can slowly mix with waters that are present in the basin of the lakes. The water quality control can be done with the aid of a combination of approaches. There are several other procedures that can be used to reduce the nutrient load concentrations. Some of them are flow control, flocculation and removal of sediments. These are considered as potential modifiers of Reservoir eutrophication.

## 8.5 OPERATION

Operational strategies always may not be beneficial even if you consider eutrophic reservoir basins. Excess drawing of water from the aerobics surface layers of the reservoir can bring about an overturn of the Reservoir water. This can bring about the bottom waters that are nutrient-rich into the euphotic zones. This will produce increased oxygen demand all over the water column and also enhance algal growth. Water-related diseases and waterborne diseases can affect human health. Most of the vectors that can produce diseases can grow in the reservoirs. Epidemics like Guinea worm disease, Bilharzia and river blindness can be associated with the construction of reservoirs. The urban settlements and industries can also populate reservoirs with waste and sediments. Some of the contaminants that are to be studied to solve human discomfort are cyanobacterial toxins, viruses, pathogenic bacteria, nitrates and organic compounds that are synthetic. Salinization is another important process that is to be mentioned while studying reservoirs. Most of the reservoirs are being installed in semi-arid and arid areas. In these areas, there will be limited water availability. Due to this, there will be mineralization or salination of the reservoirs. This problem is severe in many areas. Irrigation practices can result in the leaching of salt from the agricultural soils that may reach otherwise within no time. If there are seasonal rainfalls this problem will be increased. This will also increase evaporative concentration but the local salinities, especially in the dry seasons.

Sampling strategies are important to study reservoirs and natural lakes. Most sampling strategies must be based on the assessment programs that might bring about specific objectives of the reservoir. The linear gradation of water quality, denticity, withdrawal and off take locations can be considered as Complex factors that are to be considered as sampling strategies. The main purpose of the reservoir must be reiterated as and when the latest developments and manipulations take place. In the case of multipurpose reservoirs, we need to consider the highest water quality possible and hence there will be a need for advanced water quality measurement programs. The monitoring strategies will consider the important uses of the reservoir waters. The principal water use can be stated in the form of potable water supply, industrial water supply, power generation, irrigation supply, fisheries and Recreation, aquaculture and flood control. This site location where the primary sample will be collected is very important to frame appropriate monitoring strategies. The sampling can be done at an outlet of the supply or in its vicinity. In some cases, we can use a representative open water site. Open water sites can be used in the irrigation supply and fisheries sectors.

We can identify sampling sites at the inlet and open water sites in the case of aquaculture. To control floods, we need to consider inflow and outflow. In the case of potable water supply, the sampling frequency should be daily or weekly in continuous mode. Even in industrial water supply also there should be continuous sampling. In power generation sectors, we can monitor weekly or daily. In the case of aquaculture, fisheries, recreation, and irrigation supply sampling frequency can bi-weekly or monthly. In the case of flood control, the sampling frequency can be monthly or annually. We need to go for hydrodynamic considerations published appropriate monitoring formalities. In the case of a potable water supply, we can consider short circuit flow and thermal stratification. In the case of industrial water supply, we can opt for thermal certification. In the case of power generation, we can take into consideration internal water currents and thermal stratification. Fisheries and recreation can also be evaluated using thermal stratification. The physical and chemical properties must be considered as a part of monitoring strategies.

In this case, we need to consider nitrate, metals, organic compounds, pH, smell, suspended solids, turbidity, color, dissolved oxygen and temperature. In the case of industrial water supply, we may need to consider major ions, suspended solids and dissolved solids, hardness, pH , and temperature. In the case of the power generation sector, we need to consider dissolved oxygen, major ions, suspended solids, dissolved solids and conductivity. In irrigation supply, we need to consider the concentrations of magnesium, chloride, sodium and total dissolved solids along with pH. in the recreational fisheries sector, we need to estimate ammonia, biological oxygen demand, dissolved oxygen and suspended solids. In the aquaculture sector, it is essential to understand pesticides, ammonia, dissolved oxygen, suspended solids and temperature. In the case of flood control strategies, it is always important to take into consideration turbidity and suspended solids. Some of the biological methods not to be studied while framing monitoring strategies.

In the case of potable water supply, we need to estimate chlorophyll-a, phytoplankton species, pathogens and coliforms. In the case of industrial water supply, the pathogens might be minimal but it should be estimated accordingly. In irrigation supply, we can observe fecal coliforms. In the case of recreation and fisheries sector, we need to consider fish biomass, chlorophyll a and phytoplankton species. In the case of the aquaculture sector, we may need to estimate chlorophyll a. The objectives that were framed to develop water quality will be based on location attributes. We need to consider reservoir basins and off-take structures. The distribution of water quality variables according to the depth must be considered.

The water quality in the reservoirs will always be affected by the internal sediment pollutants and external runoff. Generally, reservoirs will drain a certain proportion of water downstream to safeguard its structural integrity. This will also ensure your sufficient power generation and water supply. This operation can be done on a seasonal scale. There is runoff, storage capacity and precipitation that can be considered as important hydrological factors that are to be considered to manage reservoirs. These factors are dynamic and can be subjected to artificial regulations along with natural alterations. This can result in altered environmental conditions inside the reservoir, pollution load and changes in hydraulic power output. All these factors can effectively damage water quality. We need to focus on hydrometeorology to specify water quality standards that might occur due to varied lots of pollution. The variation and distribution of total pollutants in the reservoirs can be affected by climate change patterns and it was observed in the Mashaba Dam reserve in Iran. The water quality parameters can be dependent on hydrological factors. The nitrite and nitrogen levels are reduced during drought periods and this is observed in several locations. If there is a flood, bulk amounts of pollutants will be flushed into the reservoir systems and this can be attributed to runoff.

Water can enter reservoirs and lakes from the rivers and streams, seepage, and precipitation. Water can be escaped from these reservoirs and lakes through streamflow discharge, subsurface seepage, and evaporation. The hydraulic residence time can be called retention time or hydraulic retention time. This is nothing but the water movement with time and we can call it as outflow and inflow of the water in the case of lakes and reservoirs. We can calculate the hydraulic residence time by dividing the total volume of the reservoir and lake by the annual flow rate. The water bodies that were classified as wide deep basins can exhibit prolonged residence times. The retention time can be defined as the time taken by the nutrients to reside in the reservoirs and lakes. We need not confuse hydraulic retention time with nutrient retention time. The recycling of the nutrients and sedimentation can affect the movement of nutrients that might occur in the reservoir system.

The hydraulic residence time can be based on several morphological factors. This can be dependent on biological and physical processes. The residence time can affect mixing, turbidity, and nutrient availability. This is the reason why the hydraulic residence time can impact the function and structure of the reservoirs and lakes. This can mainly affect the aquatic communities. This can also affect the movement of plankton, sediments, and nutrients toward downstream. Eutrophication can be considered as an important process with which most of the reservoir processes will be affected. The increased dumping of the nutrients can be considered vital to define the eutrophication rates. The eutrophication can be mainly due to vitamins, trace elements, carbon, nitrogen and phosphorous elements. Nitrogen and phosphorus can be classified as critical factors that can influence the growth of the plankton biomass in the reservoir systems. The drainage from the farmlands, sewage and industrial wastewater can comprise enhanced levels of phosphorus and nitrogen. Due to the dumping of these growth nutrients in the reservoirs, there will be increased growth of algae and green plants. The eutrophication process can be explained by life cycle theory and food chain theory.

The life cycle theory has obtained wide acceptability. This theory explains that the ecological balance will be damaged if there is an excessive discharge of phosphorus and nitrogen molecules in the water bodies. This will naturally lead to the growth of algal members in excess. The excess growth of algal members can lead to the production of algal blooms. These algal blooms can easily consume dissolved oxygen in increased amounts rendering the reservoir water with reduced dissolved oxygen levels. The food chain theory was described by Martin Shotton in the year 1997. According to this theory, there will be several food chains in the aquatic systems. This can be observed in the natural water of lakes also. It is the zooplankton that regulates algal growth. If the growth rate of the zooplankton diminishes, then the consumers of algal members will be decreased in numbers and hence exponential growth of algae. If the algal growth is beyond the consumption levels of algae, we can expect eutrophication. The organic pollutant can also result in the eutrophication process.

Water pollution can be the primary reason behind this process. Most of the sources of pollution will be the areas that are closer to the reservoir location. These sources of pollution can be classified as endogenous and exogenous sources. Some of the exogenous sources can be classified into nonpoint and point sources. Eutrophication can be high in the case of enclosed water bodies. If the depth of the water is shallow and the flow rate is slow, we can expect eutrophication. These conditions can encourage the growth of algae and plants. If the depth of the reservoirs is more, we can expect increased retention times. This can increase the number of pollutants, especially at the bottom. If the stratification starts, most of the nutrients will be liberated from the sediments and hence there will be the generation of anaerobic conditions. The ecological imbalance can also encourage eutrophication processes. This imbalance can effectively enhance eutrophication rates. The microorganisms can disintegrate the organic pollutants that are present in the reservoirs. These organic pollutants can be produced by the consumers of the food chain and producers. These microbes can be used to maintain water quality. In the case of increased accumulation of organic compounds, we can expect lowered dissolved oxygen levels. Because of this process, there will be anaerobic conditions in the reservoirs. These anaerobic conditions can damage the food chains of reservoirs and lakes. These conditions can also strengthen the Phosphorus cycles. The algal blooms can be dangerous to the aquatic ecosystems that persist in the reservoirs. The algal blooms can drastically reduce the intensity of optical radiation. Therefore, the dissolved oxygen levels will be minimized. Most of the algal members will die in absence of light and hence there will be the release of toxins from the decaying algal matter. The algal growth will be hindered due to suffocation and this will lead to disturbed ecological food chains. This will convert the phosphorus molecules that are present in water and encourages algal blooms.

The eutrophic state of the reservoirs can be based on several factors. They are transparency, chlorophyll-a, biochemical oxygen demand, total phosphorus, and total nitrogen. Biochemical oxygen demand can be taken for 5 days. We can use the lake evaluation index to know some of the parameters of water quality. Water pollution can be classified as a systematic and gradual process that can be associated with the trophic category or status of the water in the reservoirs.

Chapter

9

# Geomorphology of Rivers

## CONTENTS

9.1 Introduction.......................................................................... 150

9.2 Rivers ..................................................................................... 150

9.3 Environmental Legislation................................................... 153

9.4 Bedrock Channels................................................................. 156

## 9.1 INTRODUCTION

The large rivers can be classified as essential components of the landforms. These rivers can be always linked with the development of human civilization per se. The judicious utilization of water in the Nile valley can be treated as an example to study the importance of rivers. There are several impoundments that were formed in the last years of the 20th century. Hoover Dam can be classified as an advanced engineered structure. The quality data of the rivers can follow two main criteria. We can describe it by studying the sediment discharge rates and water discharge proportions. The river basins have occupied 70% of the terrestrial space. We can classify large rivers according to certain defined criteria. They are drainage area, average discharge, and dissolved loads at an annual rate. If the river extends to at least 2500 Km and the drainage area is about 800000 square kilometers, we can classify it as a large river. If the annual dissolved load is 100 metric tonnes and the average discharge is 7500 cubic meters/sec (Joachim et al., 1990). These rivers can be vital in sediment transfer. They can also shape the rivers to a large extent. The tectonic arrangements of the river can be very important to know. The headwaters can be present in the uplifted zones in the case of larger river systems. These rivers can flow via cratonic areas. The continental areas like Greenland and Antarctica may not allow the formation of large rivers.

In the case of New Zealand, there will be upliftment at a rapid rate. There are several active rivers in New Zealand but, there will be no formation of large rivers due to limited land. There will be the formation of uplands below the water and this can be represented by mid-ocean ridges. These mid-ocean ridges can get surfaced in areas like Iceland. Some of the river systems that are larger can flow under the sub-oceans. The large rivers can be related to the Wilson cycle that describes the production and disintegration of the continents. The tectonic setting of the rivers can be studied in relation to Cenozoic mountains, Mesozoic, early Palaeozoic and late Palaeozoic mountains, volcanic plateaus, shields, low lands, island arcs, rift valleys and plate direction movements.

## 9.2 RIVERS

The large rivers can be controlled by the balance that might exist between the gravitational spreading and plate convergence. We can observe three types of settings that might regulate the rivers. They are cratonic settings, rift settings and continental collision belts. The location and orientation of the river valleys can be linked with the structural variabilities of tectonic structures. The collision of continental belts can result in the production of the uplifted zones. There will be an amalgamation of the massive continental masses. The upliftment processes can produce drainage and catchment in the case of large rivers. If there is widespread deforestation and denudation, we can expect enlarged river valleys. Due to the collision of India-Asian plates (Figure 9.1), there was the formation of large river systems like Yangtze, Mekong, Indus, Brahmaputra, and Ganga.

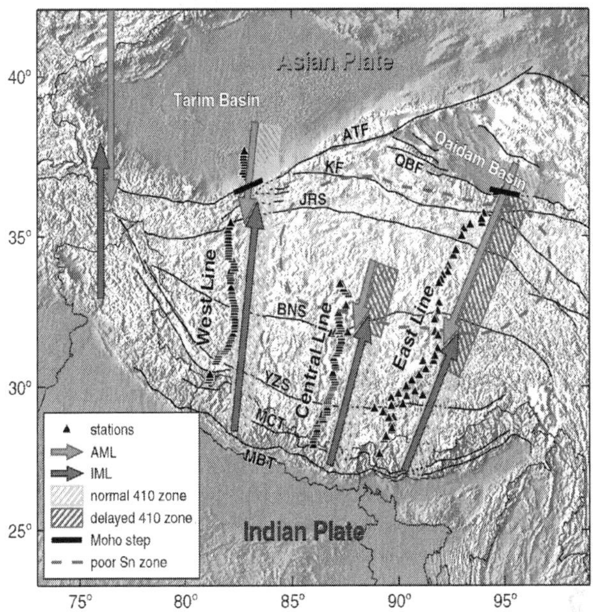


**Figure 9.1** India-Asia Plates.

**Source:** *https://www.pnas.org/content/107/25/11229*

The rivers can be developed if the plate boundaries are convergent. They can be associated with orogenic belts. This can be due to tectonic settings that comprise continental collision belts. In the case of the longitudinal foreland basin, we can expect rivers that may have headwater sustained in the mountain ranges. They may primarily form at the convergence margins. We can observe stream flow that flows longitudinally and axially to the mountain range. River Parana and Ganga can be examples of longitudinal foreland basins. We can also observe transverse systems as tectonic settings. In this case, we can observe the transverse flow. Mississippi (Figure 9.2)and Amazon rivers(Figure 9.3) can serve as examples for transverse systems.

Some rivers can be produced from the mountain belts that can be arranged in parallel to the major structures.

**Figure 9.2** Mississippi.

**Source:** *https://www.nature.org/en-us/get-involved/how-to-help/places-we-protect/upper-mississippi-river/*

**Figure 9.3** Amazon.

**Source:** *https://pisacinn.com/the-amazon-river-in-peru-and-how-it-became-part-of-one-of-the-7-wonders-of-the-world/*

These can be called intrabasinal settings. Irrawaddy, Brahmaputra, and Mekong are examples of intrabasin settings. Some river basins can be formed due to relief conditions like rift shoulders and rift tectonic settings as in the case of Rio Grande. Some river systems can be formed in case of cratonic tectonic events. We can observe this type of system in Orange and Lower Parana.

The effect of climate in the formation of rivers can be important in the regulation of the erosion process. The temperature and precipitation can be two primary climatic factors that can regulate vegetation cover. Large rivers can be formed due to increased rainfall. This is normal in the case of humid tropics. The precipitation factor can be based on differential heating, air mass movement, frontal activity, landform barriers, atmospheric disturbances, moisture, temperature, and latitudes that can define air pressures. In the case of large rivers, the precipitation levels should be adequate. Due to the increased levels of headwater regions, the rivers can yield power to initiate drainage networks and erosions. Most of the tropic latitudes can exhibit convergence of the air streams and hence we can expect enhanced precipitation rates. There are many rivers that are mainly found in the tropical zones. The mountainous regions can affect the precipitation levels and these regions can hold headwaters. These mountain barriers can bring down the moisture-packed winds and hence result in precipitation. Mostly, the precipitation can be high on the windward side. In the case of tropical areas, there will be subsidence of upper air but still, there will be the occurrence of orographic precipitation.

Deltas are the discontinuous protuberances at which rivers can make their entry into oceans, lagoons, lakes, and seas. These deltas can be subjected to redistribution of the basin processes. The deltas were classified by Galloway in 1975. This classification system was designed based on the Fisher concept of 1969. There are two major types of deltas namely high destructive deltas and high constructive deltas. In the case of the Galloway system of classification, the relative intensities of marine and fluvial processes were considered (Yang., 2013).

## 9.3 ENVIRONMENTAL LEGISLATION

The environmental legislation can be based on the level at which the environment will get strained due to anthropogenic pressure on the natural resources (Julia et al., 2013). There are several geomorphic classifications that were framed for diverse landscapes. In the case of descriptive classifications, there will be no associated with the process. They are still valuable in the analysis of the reservoirs. We can use a descriptive classification of rivers in case of inventory. By using this type of classification, we can also develop pattern-recognition tools based on GIS (Figure 9.4).

The process-based channel classifications can be used for many purposes. They can be used in the simplification of the complex processes that might occur in the landscape. This also reduces the time needed in the characterization of the basins. This classification method can be used in the stratified sampling process.

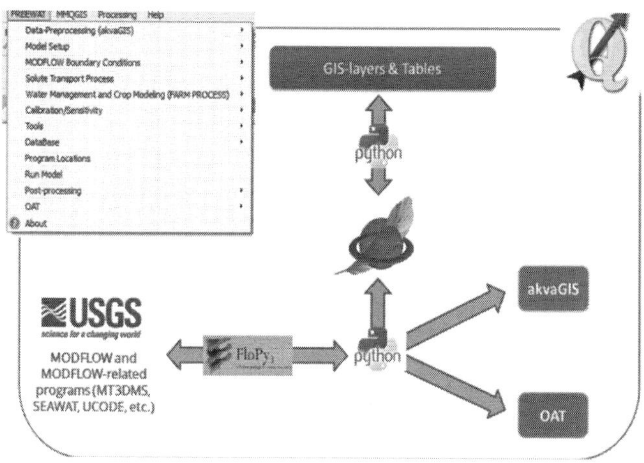


**Figure 9.4** GIS tools.

**Source:** *https://www.indiawaterportal.org/questions/freewat-training-free-and-open-source-tools-water-resource-mgmt*

We can use a limited number of samples to arrive at a meaningful conclusion. In the case of process-based classification, we can easily build conceptual models for assessing and interpreting contemporary conditions. This can be used to establish certain hypotheses. We can also formulate certain theories to forecast the anticipated disturbance to the local landscapes. We can use digital elevation models or DEMs (Figure 9.5) along with the process-based classification to know the spatial and temporal signatures of a specific area. We can easily assess how the adjacent features and phenomena can act locally and on the landscape. The interlinking patterns can be deeply visualized as and when needed. We can use this type of classification to know the channel classifications. Some of the uses are the design of streams and restoration, forecasting anthropogenic and natural disturbance of the landscapes, associating biological and physical conditions, and regulating communications. Stream order can be used to define the descriptive classification of these water bodies. The river networks can be classified into network nodes, they are tributary junctions and channel heads. We need to link these nodes with their respective positions in the river networks. The channels that are present at the tips of the river network can be called as first-order channels. These channels generally head to the junction of the first tributary. At the confluence of the two immediate first-order channels, we observe second-order channels. In general, the stream order can be linked with the physical conditions, channel size, slope, drainage area and length of the link.

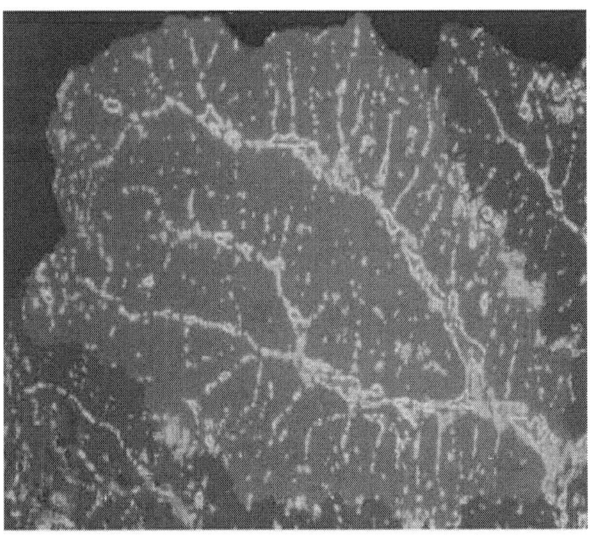

**Figure 9.5** Digital Elevation Models.

**Source:** *https://proceedings.esri.com/library/userconf/proc99/proceed/papers/pap866/p866.htm*

The spread of the river network and the ordering of the streams can be varying. The stream orders may offer us minimal information about the processes and morphology of the streams. It will give us information about the structure of the river networks. We can use this information to know the location and size of the streams that are present in the basins. We can also use structural classification to know the scales that are nested mainly in the subbasins. These subbasins can be called as hydrological units. With these hydrological units, we can easily describe the geomorphic processes of the rivers and streams (Hundey et al., 2009).

The process domains can be important. The rivers can be classified based on the sediment production rates and transfer proportions. They can be classified based on the deposition zones also. Most of the sediments can get transferred via river networks and this will occur in relation to geological time. The river classification based on the process domains can result in the formation of many channel types. We need to identify the basic geomorphic units that might be observed in the local landscape. Within these process domains, we can understand the actual behavior of the rivers and habitats that might exist within. The fluvial morphology can be studied through two important approaches. They are realistic approaches and functional approaches.

## 9.4 BEDROCK CHANNELS

Most of the bedrock channels (Figure 9.6) can easily cut into the bedrock underneath the rivers (Keith et al., 1998). The alluvial channels can be produced if there are alluvial deposits. They will get deposited on the floor via fluvial processes.

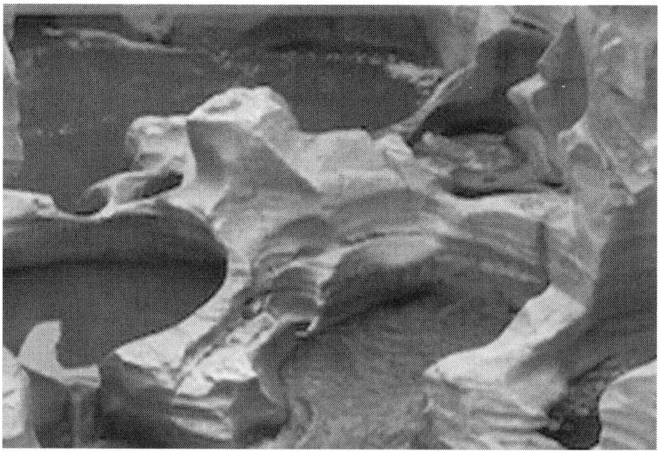

**Figure 9.6** Bedrock channels.

**Source:** *https://study.com/academy/lesson/bedrock-channels-formation-process.html*

A system can be defined as a collection of processes and objects that can associate or tie up the objects as one unit. The floodplains, channel networks and hill slopes can be linked as one unit in the fluvial systems whilst studying the sediment transport and water movement. Most of the fluvial systems can be classified as hierarchical systems. In these systems, we can observe integrated sub-systems also. These sub-systems can operate in the river basins along with other factors also. In a fluid system, the fundamental unit can be called a drainage basin. The fluvial systems can be categorized as open systems. They will transfer materials and energy into the environment. In the case of closed systems, we can witness energy being supplied to the enclosed environment only. In some cases, the surrounding environment can supply energy to these systems. The primary inputs of these systems are sediments and water that can be produced due to the cleavage of rocks underneath. Some of the other inputs are organic materials that were cleaved, rock weathering, biological materials and atmospheric inputs (Brasington et al., 2012).

Most of the energy needed for the fluvial system can be offered by atmospheric processes. These processes can condense water and also lift water. The water that was condensed can get deposited over the drainage basin in the form of precipitation. The gravitational pull can help in the movement of water down the slope. This will create energy flow through a system. The energy that is accumulated can be used in the movement of water at the sediment transport. These sediments can reach the channel network eventually. The sediment and water can get transported through a system that can lead to the outlet of the drainage basin. At this outlet, the sediment material can be discharged into the oceans. It is to be noted that all rivers may not reach the oceans instead they can supply sediments to the inlands. The loss of water can be due to the evaporation process that might occur in the atmosphere. The energy that is available can exceed the frictional forces that are related to the transport of sediments and water into the channel network and channels. Most of the energy that helps in the movement of rivers can be lost in the form of heat. Some of the material brought about by the rivers may be stored along their path of flow. Water is supposed to be stored for different lengths of time in reserved words and lakes and also in an aquifer and below the ground. Most of the sediments can be stored in the flood plains, alluvial fans, deltas, lake basins and channels. If the channel gets transported via floodplain, there will be a deposit of sediments downstream. The fluvial geomorphology constitutes three Types of systems. They are cascading systems, process response systems and morphological systems. The morphological systems can be called form systems. The landforms such as floodplains, hillslopes and the channels can produce morphological systems that can be called form systems also. Cascading systems can be called process systems. These systems represent the flow of sediments and water through a morphological system. There will be interconnected pathways that can associate hillslopes with the channels. This will be mediated by channel networks.

The process response systems can be demarcated by the interaction between two systems. There will be an adjustment between the morphological system and cascading system. This system generally represents a two-way feedback process between the form and process systems. Mostly the processes influence forms to intensity and rates of transport. The steep channels can yield increased flow velocities and hence enhanced erosion rates. The channel slope can be reduced at the steep section of the channel. Due to this reduced channel slope, the velocity of the water flow will be reduced and hence decrease erosion rates. We can understand the fluvial system clearly by dividing this into several subsystems. We can consider the system into three zones that are integrated as a whole. Each subsystem can act as a process response system. Every process response system can have inputs and outputs of its own. Every sub-system can contain certain processes that can dominate. Mostly at the production zone upstream, there will be an increased number of sediments. The sediments using the external forces can be transported through channel networks this zone can be called the transfer zone. Once the river approaches ocean regions there will be a decline in the gradients. The energy available for the transport of sediments will be greatly reduced at the deposition zone. Fine sediments can reach the oceans first. This will affect the transport of sediments into the massive ocean bodies. Coarser sediments can get deposited upstream of the channels. Some parts of sediments may reach the basin outlet.

There are several fluvial system variables that are to be considered in the field of Reservoir engineering. These variables are nothing but quantities that can vary according to time. Some other variables that are to be considered are channel depth, channel pattern, sediment yield, flow discharge, soil type, angle of hillslope and drainage density. There is a clear demarcation between external and internal variables as fluvial systems are concerned. Most of the internal variables can act inside the fluvial systems. Internal variables can be controlled by other components that exist in the fluid systems. Some of the external variables that control the fluid systems are climate and temperature. The external variables can act as independent entities and they are not supposed to be affected by the factors that control the internal variables. We need to consider human activity, tectonics, base level and climate that are external variables at the basin level. In the case of subsystems, the channel reach at the transfer zone may be considered. Some of the variables may be initiated at the exterior of the channel subsystems. In this case, there will also be internal variables that act at the basin scale. To avoid set and disparities between external and internal variables, we can call all the external variables as external basin controls.

The external basin controls can regulate the subsystems. If there is some change in these variables then it will lead to several changes and the fluvial systems. Generally, climate defines the variabilities that might occur due to different weather conditions with time. It is a known fact that weather is always changing and hence we need to take an average of the weather conditions at a particular location. We can take climate as the average of inter-annual weather variations at a particular place. Tectonics is nothing but the internal forces that can change the crust of the earth and these Limited changes can bring faulting, fracturing, tilting, warping, local subsidence and a large scale-up gift to a particular area. If there is an uplift due to a tectonic situation, the water should be fed in a lifted manner to reach higher elevations. This may need a lot of energy. If the water flows down from this uplifted area there will be increased availability of energy. This can produce a lot of sediments in most of the areas it can be attributed to the tectonic uplift process. Most of the gradients that are present in valleys may be altered or changed due to fault situations resulting in upliftment. This can also affect the design of channels. The sedimentation patterns can also be affected by tectonic movements.

The base level is an important aspect that is to be considered while studying fluvial systems. The base level is nothing but a level at which the channel is not supposed to flow or erode. Generally, the base level can be referred to as sea level. If the base level falls to the land surface, then there will be easy sediment flow and water flow into the oceans. If the base level is high then there will be limited energy available and this will yield higher deposits in the lower channel reaches. Activity can be described as anthropogenic activity that can change the course of rivers. For 5000 years human activity has changed the fluvial systems. Anthropogenic activities such as mining operations, agriculture and deforestation have altered sediment production and water flow. These anthropogenic activities can be referred to as diffused activities or indirect activities. If there is a channel engineering situation then we can assume that there is a definite change in the channel characteristics. There are several advancements in river engineering and Dam construction. These technologies have improved the construction of flood embankments, channel realignments and flood control. The channels were also analyzed using the latest technologies available today. Human activity or anthropogenic activity can be stated as an external variable and an internal variable. Some of the internal variables may exhibit autonomy and create a limited effect on the fluvial systems. Some of the variables are topography, vegetation, soils and geology. The topographic variable includes the size of the drainage basin, altitude and relief.

Most of the internal variables can be regulated by external basin controls. The dependent variables can be classified as adjustable variables whereas controlling variables can be classified as independent variables. The channel pattern is adjustable and it can be called a dependent variable. The sediment supply can be called a controllable variable and hence it is also called an independent variable. Some of the controllable variables can be adjusted by other variables that are available in the local environment. The sediment supply can be regulated by the vegetation cover of the hill slope. Element supply can be classified as an adjustable variable whereas the vegetation cover can be classified as a controllable variable. Most of the internal variables can be deemed as adjustable wearables as they are being affected by several other variables and hence, we cannot isolate countable variables from adjustable variables. It is always essential to understand the hierarchical nature that is present in the fluvial systems since all these operations matters at large scales. The climate variations can affect the vegetation cover at hillslopes and this will define the rate of sediment supply. This also affects the pattern of the channels and hence the fluid dynamics that occur in the small-scale sedimentation process in the channel can be controlled by the size of individual grains. The deposition and erosion mechanism of these grains can lead to large-scale variations with time. Time can be considered as a controlling variable and it is to be noted that every drainage basin mainly carries historical content because of the earlier events in history. There will be a combined effect of deposition, transport and erosion with time. There are several rivers that got transformed into meanders due to climatic conditions. There are some places where vegetation got recovered and hence the sediment load has decreased drastically.

Feedback systems are very important as far as the river basin studies are concerned. If a change in one variable results in a change in another variable or more than one variable then we can consider it as feedback. This feedback can reinforce the effects of change or it can also contract with the network change. There are two types of feedback that can be observed in the Reservoir engineering studies for river basin studies. Positive feedback and negative feedback are essential to study the river basin dynamics of a particular area. The positive and negative feedback can be started with one of the system variables that can be external or internal and this can lead to the chain of adjustments in a sequential manner. If they affect the original change then it is called negative if it reverses or if it improves the change it is called positive feedback. If there is a certain change in the external controls then the negative feedback and take over to stabilize systems. Positive feedback is much essential in the management of river equilibrium. Generally, soil erosion can be considered as a natural process and if the removal of soil is balanced by the addition of soil then we can expect limited problems to the environment.

The deforestation process can bring about increased soil erosion. The topsoil may contain excess amounts of organic matter that hold the soils that are rich in nutrients. Because of this, there will be increased permeability in the soils and so the rainfall will get readily infiltrated into the lower soil portions. Dependability will be decreased if the topsoil is removed and hence there will be enhanced erosion rates. The positive and negative feedback can occur in between the channels, sediment transport and water flow. These processes are quite complex to understand. The channel form can be important because it will affect the movement of sediments and water. The setting of thresholds can be an important aspect in the field of fluvial geomorphology. If the sand grains that are present on the bed are set into motion then we can assume that the threshold is crossed.

This moment can be generally restricted by the grain weight that is already submerged and also the frictional forces that are being acted on the grains. If the driving force is much greater than resisting forces that there will be no much change in the moment of water. If the driving force exceeds the submerged grain weight then there is a movement of grains of sand. So, in this case, we can assume which channel flows as an external variable. If the values are beyond the threshold, there will be an immediate change in the system. The materials that are loosely packed will become more unstable and so they slide down the slope. This can be observed in the case of landslides. In case of landslides, there will be continuous breakage of rock material and once the loose material gets dominated there will be an occurrence of a landslide that might occur in a year. Thresholds that produce a duty effect of external change can be called external thresholds. There will be the development of several instabilities with time even external changes are not dominant. Because of the several interactions between the components of the system in the fluid regions, it is often complicated to understand the responses provided by the systems. If there is a decrease in base level then there will be steepening of the channels preferable at the downstream end. Due to this, there will be an occurrence of a short break in the slope of the channel. In some cases, the short break can appear in the steep sections of the channel. Most of the erosion can occur at the cheapest portion of the channel. This can lead to the elevation of the channel. Because of this continuous lowering of the channel process, the upstream channel can become steeper gradually. the steeper zones may extend to the upstream portions of the channel. The primary responsibility of the channel to this decrease in the base level is the downcutting process. This process generally occurs at the downstream end of the channel. In this case, there will be no damage to the upper portion of the channel and therefore sedimentation trades might not be increased.

Because of the down-cutting process and its expression to the upstream, there will be a certain increase in the sedimentation rates. From the upstream, there will be increased sediments and so the downstream end of the channels receives an increase in sediments. This process can be called aggradation. Due to the aggradation process the channel gradient will be lowered. it is to be observed that this Incision zone may continue to extend upstream. Once the aggradation process gets extended upstream there will be a deposition of the sediments. The volume of sediments that is to be supplied upstream will be decreased drastically. At the end of downstream, there will be minimal sediments. This means that there is excess energy that is needed for the erosion process. The aggradation process might continue to spread at the upstream end. The increased segment supply can result in an aggradation process at the end of the channel. There will be a complex cycle that can be called episodic erosion.

The concept of scale in fluvial geomorphology is important. It will describe the processes of interactions and relationships that can occur in spatial and temporal scales. At each time scale, there will be the evolution of landscapes of different patterns. There will be several small-scale processes that control the formation of the individual grains on the bed of the channels and hence resulting in specialized landscapes. The space scale can range from millimeters to several thousands of kilometers and timescales vary from a few seconds to several years. These space scales can be classified as spatial skills. While studying the fluid systems we need to consider scale to the river basins. In the case of large drainage basin scales, we need to consider all the characteristics of drainage networks and their topography. We need to take into consideration the complete combined effect of the processes over extended time skills.

If we consider small scales, then it is essential to understand the meandering channels along with the history of the drainage basin. We also need to control the controlling variables that might occur at river channels and regulate them with time. In the case of meandering, there will be a process where interactions are relationships that might be witnessed at a small scale. We need to understand the flow hydraulics along with the bending patterns to illustrate sedimentation dynamics. There are several channel units that can be considered as depositional. These are called point bars. If you take fine-scale into consideration, it is essential to consider ripples as a separate entity on the surface of the bar. At the scales that are defined as fine, we need to consider individual segment grains.

Equilibrium can be stated as a time-dependent entity. Generally, in the case of fluvial systems, equilibrium can be defined as a balanced state that can exist in a subsystem or system. Several negative feedback mechanisms help these systems to be at equilibrium. There are different types of equilibrium that may surface with time. If you take the time domain as the minimal quantity, then it is said that the river system is at static equilibrium. Equilibrium can be classified as static equilibrium, steady-state equilibrium, dynamic equilibrium and dynamic metastable equilibrium. If we observe the same river for long periods then we can witness the ten changes that can be accountable for. There is a chance that floods matter through the channel resulting in bed scouring. When there is a gap, there will be deposits that can generate channel beds. There will be filled and scour continuously. This results in the elevation of channel beds or fluctuation of channel beds. This equilibrium can be stated as steady-state equilibrium. The elevation of channels may change according to the average conditions and this result in diminishing of channel elevation. This can be called dynamic equilibrium. There will be an effect of external Basin controlsin this equilibrium. There are only positive feedbacks in this system. There will be ascertained short-time fluctuations that occur in the river environment. This can be stated as dynamic metastable equilibrium.

The flow regime can be considered as an important aspect to be discussed in the case of fluvial geomorphology. Most of the flows that occur in the river channels can impose a lot of hydrological forces on the banks and beds. There must be a balance between the forces incurred due to flow and the boundary. The driving force of the water flow can be called an erosive force. The resisting force can be seen at the boundaries. These forces can define the tendency of rivers to be modified and hence the channel›s morphology will also be adjusted. Discharge can be considered as one of the important factors that can affect the corrosive nature of the floor. Hydrological Pathways must be studied to understand the flow generation patterns. Water can reach drainage basins through different precipitation forms. It can be dew, hail, sleet, snow and rain. The water that gets deposited on the artificial structures and vegetation can be intercepted. Some water may reach the ground and most of it will be evaporated. This can be called an interception loss. The quantum of interception can be based on different factors. We need to consider the type and extent of the vegetation. In the case of vegetation, we need to consider the actual arrangement of foliage, density, structure and size of the leaf. It is essential to consider the rainfall intensity and the wind speed (Jones, 1997). Evaporation generally occurs at the surface of the soil. We can say that it occurs at the surface layers of the soil, wetlands and lakes. Water can reach the soil surface and get penetrated the surface soil layers. This process can be called infiltration. Water that is remaining after infiltration will be allowed to flow into the down slope gradient. This can be defined as overland flow.

Infiltrated water negative transported through the soil particles and in some cases, it will be parallel to the surface of the soil. This can be called the throughflow of water. Water can also percolate into the area near the saturated zones. This can be called groundwater flow. Overland flow can be characterized by high-speed flowing waters. The speed of groundwater may not exceed 1.5 meters per day. Water can get infiltrated in the soil and this is based on local infiltration capacity. The infiltration capacity can be varied according to the soil structure in particular areas.

The weathering process constitutes the physical breakdown of rocks and chemical breakdown of large rocks into Rock materials that are loosely packed at the surface of the earth. Generally, it occurs if the materials are subjected to environmental conditions leading to instability in the surface of the earth. The weathering process can be classified as a primary source from which sediments get generated. Mass wasting and erosion are the important processes that can be witnessed in the weathering process. The weathering process can also produce mineral matter that can be used as parent materials for the formation of soils. Physical weathering process is generally observed in most places.

Some weathering reactions may involve oxygen. Most of the iron-containing minerals can easily be weathered in the presence of oxygen and this reaction can be called the Oxidation reaction. Oxygen generally gets dissolved in soil water and will be combined with minerals of the parent material. Due to this, there will be changes in the structure of crystal lattice to enhance the cleavage (Hugget, 2003). Regolith is nothing but a layer of weathered material that covers the exposed Bedrock surface. It comprises colloids, fine clay particles and large boulders. The weathering rate can be defined by organic activity, topography, parent material and climatic conditions. Mass wasting is another important process that is to be studied while describing the weathering process. It refers to the transport of rock, soil and Regolith under the gravitational force. The gravitational force can be classified into two primary components. They are slope normal components and downslope components. Generally, the down slope component can impose shear stress on the blocks under investigation. Slope normal components generally act perpendicular to the slope. We can differentiate these two components with the help of the slope angle. Increased contact time between the individual grains can enhance the rate of friction between the grains. We need to consider cohesive forces that hold the particles together. This force is much more essential if clay-sized particles and silt particles are in place. Due to the physical and chemical properties of this clay and silt, there will be the generation of electrochemical forces between the grains. There are several processes that can initiate mass moments and these processes are classified as dynamic processes. Some of these dynamic processes are earthquakes and rainfalls.

The moisture content in the soil that is already and closed by a slope can be crucial and defining the pore space is slow. Generally, water can act as a lubricant. The lubrication property can be attributed to the pore spaces between the grains. Positive pore water pressure can force the particles to exist as separate entities and hence the internal frictional forces will get reduced. Its reduction in frictional resistance can also bring down the shear strength. When these materials are exposed to sunlight are any heat source, they become dried out. In this situation, there will be minute pores that can be filled with water. This can be called as negative pore water pressure. Water can be arrested with the help of suction forces.

The negative pore water pressure can also bring about increased shear strength. This can work well with silt-sized particles and clay particles. Most of the landslides that are present on the soil surface can be vulnerable to storms. Pour water pressure can be classified as a dynamic entity that can get altered with time. There are different types of mass movements. Some mass movements may occur for a short time and this can be witnessed with mass movements like solifluction, creep and heave. In the case of slides, there will be a shared plane between slope and moving material and this is two-dimensional. Flows can be differentiated from slides as there will be no share plane. Mostly the shear can occur at the lower portion of the flow. Generally, the liquid content that constitutes the flow can differ. The mudflows can represent fluid states but with half liquid and half solid states. Solifluction can be classified as the slow flow type.

The transfer of sediments from the hill slopes to the channels can be studied by different methods. Primary erosion can be used to define the in-situ erosion pattern of the rocks, soils and Regolith. There is an important relation between the sediments that come from hill slopes and the erosion of the sediments. This transfer is dependent on hillslope channel processes. This process can be named as the hillslope channel coupling process. The sedimentation transfer that may happen at channels can be much limited in the case of decoupled systems. The size of the sediment can determine the extent of coupling. Finer sediments can be classified as mobile sediments. This slope channel coupling can be stronger if the sediments are fine.

The water flow in the river can be regulated by two important climate-related processes. They are evapotranspiration and precipitation. River discharge can be calculated by subtracting evapotranspiration from precipitation. Water flow in the streams can help human populations to a larger extent. Most of the terrestrial landscape is formed due to fluvial processes. It is always essential to study geomorphology in detail to understand fluvial hydraulics (Figure 9.7).

**Figure 9.7** Fluvial Hydraulics.

**Source:** *https://www.grk802.tu-braunschweig.de/Mitglieder/GiovanniMichelazzo.htm*

The fluvial hydraulics can describe the transport of water via a stream network. Streams can be any flowing water body that exhibits measurable velocity. There are no specific quantitative boundaries that explain streams. Stream reach can be any segment that is uniform in shape and size with well-defined flow characteristics and channel materials. Every stream has specialized forms that can be defined by sediments and water flow. These forms can be contributed by the closest drainage basin. We need to consider climate settings, topography and geological patterns. Natural streams can have complex and dynamic properties. We need to understand watershed to know the fluid hydraulics in detail. Watersheds can be called a catchment or drainage basins.

It is nothing but an area that can hold all the water that may be deposited on the landscape. Any surface that can demarcate watersheds can be called a divide. Drainage of the land surfaces can be possible through stream networks. Most of the steam reaches are enclosed in the networks. The network pattern can be associated with spatial rearrangement of channels at the terrestrial surface. The spatial rearrangement can be brought about through geological structures and land slopes. Most drainage networks can follow dendritic patterns. There are certain problems with open Channel Hydraulics. In order to understand flow channel Hydraulics, it is always important to design canals and channels. We need to consider the extent of flooding in case of snowmelt and storms. The rate at which the flood will still get transported to the channel network must be estimated. The quantity of the sedimentation and the size of sediments must be evaluated before defining the properties of open channel flows. We need to categorize the flow resistance patterns. We need to estimate the magnitude of the flow. There are some methods that can be used in understanding fluvial morphology. Some of the thought methods are inductive method and hypothetico-deductive method. In the case of the inductive method, there will be a collection of data from the field, literature, and laboratory. These are based on the hypotheses. In the case of hypothetico-deductive methods, we can use modified conceptual models. We can use experimental and comparative methods. The comparative and systemic methods can be used. We can use several observations made simultaneously at several reaches and multiple rivers. There will be several levels of drainage networks. This will be based on potential thresholds, sensitivity to variations, functioning types. We need to consider diverse spatial objects like observed and control. We can use associated techniques and working methods that can be developed through inductive methods. It is always important to use preliminary hypotheses that can be tested.

We can use experimental methods also. There must be certain controlled conditions that are to be maintained in the lab. In fields, it is not possible to maintain constant conditions. These approaches can be dependent on specialized frameworks. The experimental techniques can be considered as important among all these methods.

We can use pre-field methods, field methods, laboratory methods and post-field methods. The pre-field methods can constitute thought methods and they can be used to design data collection. We need to know the sampling protocol. In case of field methods, we need to measure the processes that can be involved in the deposition process. In case of field methods, we need to consider archival data and spatial information that is needed. In the case of laboratory methods, we need to use the lab in order to analyze the biotic, chemical, and physical methods. When it comes to post-field methods, we need to opt for data interpretation and analysis. There are five kinds of geomorphological investigations that are to be questioned. The historical framework must be considered while understanding the fluvial geomorphology and hydraulics. We need to know and apply the historical trend of the geomorphological variables. The spatial framework must be studied. The nested character and spatial structure must be described in advance to reach a meaningful conclusion. We need to consider biological, physical, and chemical methods. We need to analyze the forms and processes that are associated with fluid hydraulics. It can be done by measuring the water flow and sedimentation rates. We need to assume future frameworks also. There are several techniques that can be used in modeling, simulating hydraulics.

Most of the environmental scientists will be having a proper understanding of the streams and rivers. Most of the riverine environments will be subjected to several environmental stresses. There will be chemical and physical hazards that can alter the river course. The geomorphology has become complicated in relation to the fluvial systems. Most of the geological information can be linked with the river deposits and in this case, we need to know the reality-induced approaches. There will be some deposits that might be deposited due to the action of rivers. In this aspect, we need to consider surficial geology. This refers to the study of rocks that can be classified as unconsolidated materials. These materials can be observed close to the land surface. The alluvium deposits can occur at the landscape in the bar deposits and modern channels. Most of these deposits will be associated with the alluvial fans, terraces, and floodplains.

There are two facies' assemblages. They can be classified as a top stratum and bottom stratum. Some of the lithofacies can be variably laminated with clay and organic silt. There will be channel fills that are abandoned. In the top stratum, we can see bedded sand, silt and clay and it will be massive. Some of the environmental facilities are splay, levee, splay point bar with backswamp with levees. At the bottom stratum, we can find sorted cobbles, gravel, and sand. We can also find the cross-bedding patterns made up of sands. There will be point bars and channels as the environmental facies. The sedimentation unit that can be classified with age can be sampled randomly.

There should be an estimation of the particle size distribution that is mostly unbiased. We need to consider sedimentary features. This can be classified as a time-independent approach. These age equivalent units can be classified into several facies' units. This can be done prior to the random sampling process. That can be random sampling and stratified sampling. Due to this, there will be increased efficiency and hence total variance can result. We need to amp facies and ages with the field data that is available.

We need to select the representative samples that can be having normal facies to all the units. The logistic constraints and costs that are incurred whilst field studies must be considered. In the case of sediment studies, we need to consider the provenance of the sediments as an important aspect. We need to consider characteristics of particle size, chemical, mineralogical, and lithological importance. The mineralogical and chemical features pertaining to the sediments should be considered while evaluating the routing of the sediments. There will be aspects that pertain to the contaminated sediments. We need to consider geomorphic, sedimentological, stratigraphic approaches in order to understand the previous river flow patterns. Mostly, these approaches can provide us certain evidence about the past morphology of the channels. The architecture of the facies will be studied to know the flow conditions. We can use the paleo hydraulic tools to know the hazard, paleoclimatic issues and others.

The regime-based approach can be used to know the empirical relations between the past flow conditions, sedimentological features, and channel morphology studies. We can also use the flow competence methodologies. The theoretical and empirical relations can be considered to evaluate the size of the clasts and the strength of flow. The record of the individual flood events can be stored in the slack water deposits. This can provide us an idea about the previous flood stage and events thereof. Some of the important attributes that are to be considered are river type, analysis scale, retrodicted properties and other applications. The regime-based methods can be used to understand the previous flow conditions. The geochronology of alluvium deposits can be considered an important aspect. Geochronology can help us understand the ages through which the alluvial strata will be developed. There are certain methods that can be useful to know the alluvial sediments and late quaternary sediments. We can use dating methods. These dating methods can be classified into relative, numerical and hybrid methods. The relative dating process can be based on the stratigraphic superposition, weathering features and morphological features. In the case of stratigraphic superposition, we can easily determine the deposition sequence. This can be classified as a highly reliable method. This method needs deep knowledge about the trenching processes, drilling, and other associated methods. While considering weathering characteristics, we can know the age trends of the soils. We can also know the weathering rinds. Whilst considering morphology, we need to consider the elevations that might be present in the alluvial plains or terraces. This data can be used to define the deposition sequence of the fluvial systems. It is always important to know the mutual relationship between allostratigraphic and terrace units. In case of major events, it is always important to analyze the regional correlations. We need to know the numerical dates and weathering chronology. The erosion degrees of the terraces can represent the ages of the fluvial systems.

In the case of numerical dating, we need to consider radiocarbon. We can use radiometric methods to know the age of the alluvial deposits. It is always essential to perform sampling and careful processing must be done to lower the errors. The interpretation mainly encompasses wood, charcoal, leaf litter and organic fractions of soils. We can also understand the carbon levels that are being derived for ages. The numerical dating can be done using photoluminescence methods. This method can be useful in dating the artifacts and sediments. We can use sunlight and heat in this process. The techniques like optically stimulated luminescence and thermally stimulated luminescence can be used. This method can be reliable and can offer us precision in dating techniques. Radioactive lead can be used in dating methods. Generally, lead can be produced in the atmosphere. This element can be absorbed by particulates and precipitation. The half-life of this element can be 22.3 years. We can use cosmogenic isotopes in the dating process. In this method, we can use calcium, carbon, neon, helium, chlorine, aluminum and Be as materials for the dating process. In this method, we need to consider mass spectrometry and chemical extraction. Dendrochronology can be used as an important technique. In this technique, we can use tree rings to know the sedimentation rates.

There are some hybrid-correlative techniques that can be used in the dating process. Pedology is a study that deals with soil formation. Several biological, chemical and physical processes can convert the alluvial sediments into soil profiles. In the case of fluvial geomorphology, we need to consider time factors. We can use pedogenic features in dealing with the dating methods. The pedogenic features can be used in the interpretation of environmental change and the environment. The aluminum and iron oxides can be studied to know the secondary accumulations. The changes in these oxides can be used to record the variations in the soil drainage that may be changed according to the time. The combined soil profiles can be used to know the actual sedimentation process. The deposition of the sediments and soil formation can bring about the cumulative soil profiles. These profiles can explain the sedimentation rates and horizon differentiation thereof. Most of the sedimentation rates pertaining to the alluvial sediments can be considered as a rapid process. We need to know the stratigraphic distribution and spatial distribution of the soil profiles. These profiles can describe the changes that might occur in the river with time. The pedogenesis can be measured with the occurrence of the lithofacies. We can use this information to determine the pedofacies. The pedogenic attributes can be important to describe the fluvial geomorphology. Using the pedofacies units, we can describe the sedimentation rates of flood plains and channels.

Stratigraphy is another important technique that can be used to know the sequence of soil formation. We can also understand the spatial frameworks of the geologic column. In stratigraphy, we can integrate pedology and sedimentology to understand the alluvial records. Allostratigraphic units can be used in fluvial science. This unit can be defined as the stratiform body that can be generated from the sedimentary rock and it will be delimited by the discontinuities. The allostratigraphic units can be used to explain the alluvial deposits. It can be used to know the upper boundary of the geomorphic structures. These allostratigraphic units can be classified into facies. These facies can be associated with the weathering process, age, sedimentological factors, particle size, minerals and lithology. Mostly, these allostratigraphic units may be classified as diachronous. In some cases, we can expect isochronous nature. The surface geological data must be collected to know the actual characteristics of the ground surface. Most of the data that is needed can be based on time and money. There are many types of data that can be collected from the subsurface. The river segments and reaches can be exposed naturally at the floodplains and terraces. They can be observed at the cutbacks also. The land use variations can affect the sediment supply and runoff. These variations can change the ecosystems and riverine environments.

Archaeology and geomorphology can be considered as inter-linked concepts. These two subjects flourished in the 18th and 19th centuries. The artifacts can provide geomorphologists a reference point to identify the exact data of a geologic process. They can be used to date the rocks with the help of inscriptions. It is always important to maintain the records of the landscape for several years as these records will help us understand the changes that might have occurred in the landscape. The stream width and channel migration can be studied using land surveys. The channel cutoff information can be achieved by the travel accounts, surveys, and botanical evidence. The information pertaining to the cross-section changes can be collected from the repeat channel surveys, discharge measurement records and stage-discharge ratings. We can also use bridge surveys. The channel change information can be obtained by the historical ground photos, location of buildings and level of the buildings.

The information about the channel incision can be collected from the topographic records. We can obtain the bedform profile and long profile using general navigation surveys. Travel accounts can be used to know the information about the fallen trees and distribution, nature of the woods. We can use artifacts that are datable to know the information about the sedimentation process. We can use reservoir storage variations and lake sediment chronologies to know the information pertaining to the sediment yields. The slackwater deposits can be studied to know the palaeo floods information.

The botanical evidence can be used to know the flood dates. The flood records obtained from the buildings and water levels can describe the hydrological conditions that exist in the local area. We can also use newspapers, dairies, and journals to know the information about the hydrological conditions that existed in a specific area. We can also use log books, newspapers, and diaries to get information about climate change. There will be problems with accuracy as far as the maps are concerned. There will be several generalizations and abstractions that mimic reality. It is always important to prevent mistakes that might get surfaced during the analysis of the fluvial geomorphology. We need to consider positional accuracy also. Generally, positional accuracy can be dependent on technical limitations. Most of the survey equipment may provide us good accuracy rates but during their operation, there will be certain other external factors that can disturb the accuracy levels. Attribute accuracy can be important and it deals with the attributes that are being mapped. We need to use scale that is mutually acceptable with the agencies to know the actual extent of geological structures. The positional accuracy can be evaluated by comparing the measure of true position with the mapped positions. This can produce certain displacements that need to be analyzed with the random error components and bias. We can eliminate the geometric transformations.

The systematic approaches are essential in studying fluvial geomorphological studies. The system concept can be employed in environmental science for almost all studies. It involves biotic processes, chemical and physical processes. The system concept can be used as an important tool to process research in meaningful dimensions. The system can be considered as a combination of many components that can be integrated. There will be the transfer of matter and energy between the system and the surroundings. The fluvial systems can be classified as an adaptive process system. They can be classified as complex systems. There are 2 primary components. They are cascading systems and morphologic systems. The cascading system can involve sediment and water whereas the morphologic systems may involve deltas, hill-slopes, floodplains and channels. Most of the fluvial systems can get changed with geological time. This can be attributed to depositional and erosional processes. In most cases, the fluvial systems can be subjected to human impacts, tectonics, base level and climatic conditions.

Most of the zones that are associated with the fluvial systems can be classified as open systems. Each portion of the attributes pertaining to fluvial morphology can be linked with sediment movement and water discharge. If we considered the lateral dimension in the fluvial systems, we need to consider the bidirectional links between the margins of the river and the main channels. The alluvial corridors must be treated as complex entities. There will be certain paleo forms that consist of screes, alluvial fans, and terraces. Most of these components can change the channel features. The geological nature of the landscape can affect the width and slope of the valley floors.

The hydro systems can be classified as the fluvial system that is extended. This concept can be used for most rivers. These rivers can have developed floodplains. We need to consider the biological and chemical aspects of the geomorphological parameters. The fluvial systems can illustrate both longitudinal and vertical dimensions. The floodplains that are large can affect the engineering structures. This can be related to the channel forms and natural processes. The fluvial systems and associated components can be studied at different spatial scales. The objective of the observer can be important and this determines if we need information in lesser detail or greater detail. The dendritic drainage pattern and large segment can be important in determining the local geology patterns. The fluvial systems can be distinguished by the asymmetry that encompasses different scale levels. Mostly the basin-scale can affect the reach scales. The dams can create incision downstream and exhumation of the boulders. There will be the simplification of the channel networks. There are several components of the fluvial systems that can be studied at several spatial scales. It is always essential to the components that deal with geomorphology, geology, hydraulics, and hydrology at all scales possible. There are several interactions that might exist between the zones. These zones can be illustrated based on climate, relief and development.

Geological conditions like structure and lithology can be considered as important in describing fluvial systems. The drainage density can be represented by D. Stream gradient (S), discharge per unit area (Q), Peak discharge ($Q_p$), sediment yield ($Q_s$), the ratio of channel length to the length of valleys (P), must be considered. In dry conditions, at the production drainage basin, there will be increased D and S. The cascading systems in this case can exhibit high Q, $Q_p$ and $Q_s$ but low $Q_b$. The transport river can exhibit a bed load channel with low P, high width to depth ratio and high S. We can also observe flashy flow, high bed load and high $Q_s$. We can also obtain the deposition of the piedmont coast in case of dry conditions. At these zones, we can obtain high sand to body ratio, fan delta, Bajada, and alluvial fans. There will be several discontinuities and rapid deposition.

In the case of wet morphologic systems, there will be low S and D. The cascading systems at the production drainage basin can exhibit Low $Q_s$,$Q_p$ and Q but high $Q_b$. We can expect suspended load channels with high P but low width/depth and S. In the case of cascading systems, there will be steady flow with low Q and bed load. There will be a minimal sand body ratio with alluvial plains and deltas. There will be steady and slow deposition. The analysis of the river systems can be dependent on the spatial objects that can be related to the system components. The fluvial systems can exhibit interacting components and open components. Some of the interacting components can be classified as nested.

The similarity analysis can be used to study the landforms that exist at a spatial scale. This can be used to compare the attributes. These attributes can be used to demarcate the groups that determine the fluvial patterns. We can use this system in the analysis of the partial systems. There will be several similarities that need to be evaluated at the spatial scales. We need to define certain channels that can be used to analyze the similarities. The similarities can be present in the cross-sections of these entities. This analysis can be classified as synchronous. This is due to the acquisition of several measurements that are made at the same time. We need to do the statistical analysis later. We can follow connectivity analysis. This can be associated with the components that are present in the fluvial systems. They can be former channels, channel reach and basin channels. This analysis focuses on the components and logical units. The retrospective analysis and diachronic analysis can be used to assess the variations or fluctuations that may surface with time. We can use historical sources in this matter. We need to consider the chronology, spatial extent, and intensity of the variations. With these analyses, it is easy to assess the dependable variables. In the case of diachronic analysis, we can easily notice the adjustments that might occur in the system components. These analyses can be used to know the land use changes that can be both dependent and independent on the watersheds. The analyses can be useful in understanding the coastal erosion processes.

The systems approach can be classified as flexible and it can be qualitative and quantitative. The system concept generally utilizes the preliminary hypotheses if needed. There will be an established framework that can accumulate or combine the geomorphic tools.

There are two approaches that can be used in geomorphic investigations. We need to use the location of the time substitution techniques to initiate the evolutionary models of the land-use variations. We need to consider the condition evaluation for assessment of the landform sensitivities. The stability of the landforms should be considered accordingly. In the pre-restoration survey, we need to measure the geomorphic processes and measurements in the restoration operations. This step can be followed by the project implementation stage. In the case of a post-restoration survey, we may consider the geomorphological process again.

The LTS tool can be considered as a popular tool and this can be called an ergodic method. In the case of LTS, we need to compare the features that can be generated by the similar processes that are being operated at similar physical conditions. The incised channels that are formed with the alluvium can be defined by analyzing the cross-sections of important locations using surveying. Some of the observed sequences can represent the temporal evolution of the special features. The location of the condition evaluation is another method that is being called as LCE by geomorphologists all over the world. This evaluation can measure the features pertaining to the landforms that are unstable and in some cases they can be stable.

Chapter

# 10 River Ecosystems

## CONTENTS

10.1 Introduction .......... 178
10.2 Corridors .......... 178
10.3 River Corridors .......... 180
10.4 Aquatic Ecosystems .......... 181
10.5 Riparian Ecosystems .......... 182
10.6 Ecological Functions .......... 184
10.7 Habitat Fragmentation .......... 188

## 10.1 INTRODUCTION

Most of the dam projects can be classified into storage projects and run-of-rivers. Generally, the ecosystems can be of different sizes. We need to understand the ecosystems in detail to know the information about the streams, floodplains, watersheds, and landscapes. There will be an external environment and an internal environment in the ecosystems (Jens et al., 2019). These environments can be represented by energy transformations like cooling and heating of the stream water. The movement of the insects, fish, and mammals along with the materials can be related to these environments. There are many sub-ecosystems that can result in the generation of river ecosystems. These river systems can be a component of the major landscape ecosystem that can be close to the actual riverine environment or may be remote but linked with other small lakes. The structure of the landscape ecosystems and their function can be defined by the flow of the water bodies and their deposition and erosion characteristics. The river ecosystems can have output and input relations with the nearby ecosystems that constitute the landscape. If we want to restore the river ecosystems, it is always important to know the relationships between the components of the ecosystems. We need to consider four important components that can define the spatial structure of the riverine ecosystems. They are matrix, patch, corridor, and mosaic. In the case of the matrix, we can observe land cover that can be interconnected with the territorial surface. The matrix can be assumed as any land cover type. We can find adequate forests and agriculture spread over the landscapes. A patch is nothing but a polygon that is a nonlinear area. It can be less abundant. We can differentiate patches from matrices. Another important spatial component is the corridor. It can be classified as a special patch. It can associate all the patches present in the matrix. The corridor can be elongated or linear in shape. These elongated corridors can be called a stream corridor. In the case of mosaic, we can consider the accumulation of the patches. These patches may be completely dominant. They may not completely connect the landscape.

## 10.2 CORRIDORS

The stream corridor (Figure 10.1) can be a specialized ecosystem that may constitute the external and internal environments (Ellen et al., 2019). The external environments can include landscapes that are present in the vicinity. Most of the stream corridors can act as primary and fundamental pathways to carry organisms, materials, and energy. The inflow and outflow of the entities can be defined by the spatial structure of the native landscapes. The spatial structures can be modified with time. We need to know the feedback loops that can exist in the structure and movement. There are several functions that can define the stream corridors that can be associated with the drainage networks. We can call the complete parcel of the landmass that can be interlinked with several components of water resources and collection can be called a watershed scale. Watersheds can be studied at various scales. The corridor, patch and matrix can be ambiguous terms that can be used to define the ecological structures that might exist in the watershed. We need to classify the watersheds into lower, middle, and upper watershed zones.

**Figure 10.1** Stream corridors.

**Source:** *https://en.wikipedia.org/wiki/Stream_restoration*

## 10.3 RIVER CORRIDORS

River corridor (Figure 10.2) can be considered as an important spatial element that can be present in the watershed scale (Bertuzzo et al., 2007). These corridors can be present in the landscape scales too. There will be matrices that can be in the stream corridors. These stream corridors can constitute shrub cover, herbaceous vegetation, and riparian forests. There will be the existence of some patches at the wetlands, residential developments, oxbow lakes, grassland patches, shrub lands and forests in the channels.

**Figure 10.2** River corridors.

**Source:** *https://floodready.vermont.gov/flood_protection/river_corridors_floodplains/river_corridors*

Generally, the river corridors can be classified into plant communities and structural features. Some of the riparian zones can exhibit both characters. Most of the riparian zones can be classified as transitional. This transition zone encompasses upland and wetlands. The biological communities can play a vital role in framing the landscapes. These communities can be present in aquatic ecosystems and terrestrial ecosystems. The plant communities can be readily under the terrestrial ecosystems. The vertebrates like amphibians, reptiles, fish, invertebrates, and aquatic plants can be classified under aquatic ecosystems. The terrestrial ecosystems can include birds, mammals, reptiles, amphibians, and plant communities. In stream ecology, we need to consider the plant communities as fundamental components. The ecological integrity can be defined by the relationships that might exist between the plant communities and ecological components. Most of the biological communities can depend on plant life for energy needs. Most of the plants can moderate the solar fluxes that might reach the aquatic habitat. It also affects the physical habitat of the aquatic ecosystems to support the aquatic biological species. We can observe increased vegetation on the slopes. This slope region that supports dense vegetation can be classified as primary product that is needed by the aquatic ecosystems.

The composition of fauna can be based on increased connectivity between the patches of vegetation, seasonal migration routes, vertical and horizontal diversity of the habitats, critical microclimates, cover types based on the spatial and temporal constraints, food availability, higher primary productivity, and availability of permanent water. The disturbance in the river ecosystems can be brought about by channel migration and flooding. Moist environments can bring about enhanced primary productivity. It also affects the biomass of the riparian ecosystems. The stream corridors can result in moisture extremes, temperature variations, evapotranspiration, shade, and water fluctuations.

## 10.4 AQUATIC ECOSYSTEMS

The aquatic ecosystem can be considered as an important aspect of the landscape. The stream biota can be classified into vertebrates, macroinvertebrates, micro-invertebrates, protists, macrophytes, algae and bacteria (Stefan et al., 2003). The vertebrate's species may include mammals, reptiles, amphibians, and fishes. The macro-invertebrates include snails, clams, worms, crayfish, etc. The micro-vertebrates include nematodes, ostracods, copepods, and rotifers. The protists include ciliates, flagellates, and amoebas. The macrophytes are nothing but the higher plants.

The aquatic plants can comprise mosses. These mosses can be attributed to the substrates completely. There will be certain aquatic vegetations that are rooted in the aquatic ecosystems. Due to increased flow rates, these rooted plants will lose the substratum. There will be increased growth of the vascular plants in certain regions. This can be common in the case of slow-flowing water, high nutrient loads, stable substrates, and water purity. The leaf litter can be bulk and it can be cleaved into minute particles. These particles can be absorbed by the invertebrates. Amphipods and insect larvae can be classified as shredders of the aquatic ecosystems. The caddisfly larvae can be named as filterers. The fishes that are present in the aquatic ecosystem can be called apex predators. The habitats can be affected by the organisms in the streams and biological diversity. The stream system can function in a stable state. The habitats and diversity can be encouraged. The cross-section that step walled and narrow can provide a limited habitat that can be used by the biological communities. The habitats can get increased because of the sinuosity of the streams. If the sand bed is uniform, then there will be limited habitat diversity. The pool riffle, step pool, rapids, boulder cascades and debris dams can encourage biological organisms in the streams. A river can be assumed as an individual entity that can be functional. The vital role of the plants is to regulate wildlife and fish growth. If we fragment the natural landscapes, then the natural cover types will be disturbed and hence negatively impact the wildlife and vegetation. If there are minute fractures in the corridors, the movement of animals will be negatively affected. The ecological conditions can be determined by the flow conditions of the water. The streams can be demarcated by the flow of water that can be downstream or upstream. The microspecies and other species can be affected by the spatio-temporal changes in the flow. Most of the flow velocities can regulate the nutrients and foods that are needed for the organisms. The flow velocities can also remove these nutrients from the sites that are classified as primary supporting sites.

Most of the streams can exhibit flow rates with limited mobility. The fauna that is present in the bed and on the banks might exhibit the same configuration and composition. The increased flow rates can help in scheduling the spawning and migration of fishes. Fish can observe high flows and they can migrate easily. The temperature of the water can be important in determining the properties of the stream and in comparing other streams. The temperature factor can be important in the case of cold aquatic animals. Most of the biochemical and physiological processes will be affected by variations in temperature. The stream insects can develop and grow with the increased temperatures at certain portions of the streams. This can be variable with the seasons. There are some species that can complete their life cycles at warm sites rather than cool sites. We can observe temperature-dependent growth rates in fishes and algae.

## 10.5 RIPARIAN ECOSYSTEMS

The diminished temperature and light can alter riparian vegetations. If the water flow is slow, the sunlight can easily heat the water and this is evident in the summer season (Ronald et al., 1982). This condition was observed in Pennsylvania when the flow rates dropped in the summer season resulting in a rise of 12°C. In the winter season, there will be a reversal of these conditions that can lead to lowered temperatures. The riparian forests may have buffers that can avoid temperature fluctuations. These buffers can mitigate the temperatures that can be increased due to deforestation. Oxygen can be considered as an important abiotic factor that can be considered in riverine ecosystems. Most of the oxygen will be absorbed by the water body from the atmosphere. This can be done via photosynthesis. Most of the mountain streams may not get the waste discharges. These mountain streams will be filled with oxygen. This can be due to constant exposure of the land surface, constant motion and depth that is generally shallow. Most aquatic species can be dependent on the dissolved oxygen levels. If the dissolved oxygen is sufficient, we can expect the luxuriant growth of the aquatic species. This also can bring about vigor among the aquatic species. In case of limited dissolved oxygen levels, most aquatic species can experience stress and some may not compete well with the other species that might be sharing the same food resources. The dissolved oxygen concentration can interfere with the fish's growth. If the dissolved oxygen falls below 3mg/L, we can expect lowered fish populations. If the dissolved oxygen exceeds the limits, most of the fish will die. The concentration of the dissolved oxygen may be lowered if the flow rates are lowered. The enhanced aquatic matter concentrations, algal blooms and rooted aquatic plants will be increased due to slow flow rates.

The increased pollution levels can also reduce the oxygen levels in the streams. The main factors that can define the oxygen levels in stream waters are salinity, pressure, temperature, and the number of aquatic plants that can survive in the water bodies. Fish can luxuriantly if the dissolved oxygen level exceeds 5g/L. pH values can be considered along with other factors. Most aquatic life can survive at neutral pH. This means the water body has H ion activity. In case of pH fluctuations, water in the stream can become alkaline or acidic. Some of the soils can change the pH of the buffer. The substrate can be considered as an important component in aquatic ecosystems. Most of the substrates that are present in the aquatic environments can affect the biotic communities. Some of the substrates can be produced at the water-substrate boundaries. This zone can be called a hyporheic zone. At this zone, there will be luxuriant growth of benthic species. They can easily grow and reproduce in this zone. This zone can be about one meter thick or even less. This zone can be a major subsurface environment. The substrates present in the streams can comprise woody debris, organic matter, boulders, cobbles, gravel, sand,

and clay. Some of the substrates can alter the interstitial flow patterns and can modify structures. These substrates can also affect the deposition of organic materials. Due to this, there will be decomposition and production processes. Slit and sand can be classified as minimal substrates that can support the growth of the aquatic species. There will be limited species that can live in the silt and sand. The eutrophication rates and nutrients present in the aquatic ecosystems can also define the growth rates of the biological communities present. silica, selenium, potassium, phosphorus, and nitrogen are essential in plant growth. Phosphorus and Nitrogen will be needed by the plants in heavy concentrations. These elements can also bring about excessive growth of aquatic flora and algae in the stream ecosystems. We can define this process as eutrophication.

## 10.6 ECOLOGICAL FUNCTIONS

The ecological functions (Figure 10.3) of the river ecosystem can be important as they define all the essential features of the ecosystem (Manfred et al., 2013). Some of the basic ecological functions are sink, source, barrier, filter, conduit, and habitat. It is always important to opt for restoration of local ecological functions as and when needed. Connectivity and width can be classified as two fundamental features of the river ecosystems.

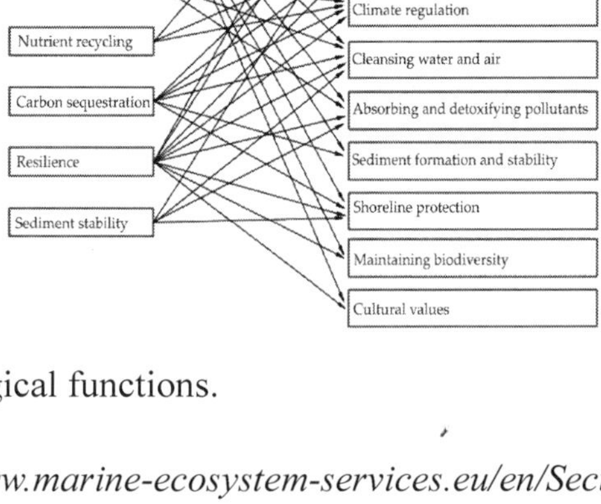


**Figure 10.3** Ecological functions.

**Source:** *https://www.marine-ecosystem-services.eu/en/Section-3/3-1-building-a-common-language-definitions-and-nomenclatures/3-1-2-do-natural-habitat-ecological-functions-and-ecosystem-services-always-match*

The connectivity features can be defined by the minor breaks that intervene in the corridors. These breaks can form between the adjacent land uses and streams. The movement of fauna and flora and transport of energy can be affected by the connectivity component of the streams. The width can be considered as an important aspect that is to be considered. In the case of stream corridors, we need to consider the distance that is in between the vegetation cover and stream to define the width of the streams. The width can be regulated by anthropogenic activity, ecosystems, environmental disturbances and gradients, community composition and edges. The different habitat needs and average dimensions can bring about considerable changes in the river ecosystems. These features can be attributed to the width component. Habitat function must be studied to understand the riverine ecosystems. The area that can support animals and plants can be classified as habitat. In these habitats, the living organisms can reproduce, feed, grow and live for many years of their life cycle. The stable biological communities can be evaluated by the genetic variations, number of species and size of the population. The streams present in the ecosystems can define the actual nature of the landscape. Stream corridors can support appropriate habitats that can be wider and may exhibit enhanced connectivity. Variations in animal and plant communities can change the morphology of the river valleys. The diverse streams and broader water bodies can support most of the species that can be confined in some cases. Precipitation, solar radiation, and soil wetness can change gradually and this can influence the river ecosystems in several ways.

The conduit function can also affect the river ecosystems. The conduits can serve as a route that can allow the organisms, materials, and energy. Stream is nothing but a conduit that can easily transport sediment and water. There will be moving with the stream and from the adjacent paths. The stream corridor, aquatic fauna and sediments can be transported alongside the stream corridor. The corridors can be classified as longitudinal and lateral conduit functions. Most of the stream corridors can be enclosed by canopy, mammals, and birds. This can be possible with the streams that can pass through the vegetation.

The movement of nutrients and organic debris can supply food to invertebrates and fishes. Most of the corridors can serve as habitats and pathways. They can be suitable for mobile wildlife and migratory birds. Most of the stream corridors can be used as filters. This will aid in the selective permeation of organisms, materials, and energy. This can lower water pollution and can reduce the transportation of the sediments. These corridors can act as natural boundaries to wild species, plant communities and land uses. There are several organisms, energy and materials that can be transported perpendicularly to the streamflow. Some of the elements can also get transported parallel to the stream corridors. The movement of water on the land, sediments and nutrients can be effectively filtered by the riparian vegetation. Some of the dissolved substances of nutrients, phosphorus and nitrogen can enter the river valley easily. Some of these substances might not enter the channels easily but can get confined to soil organic matter, clay, root absorption and friction. The edges can be present at the intersections of the stream corridors that can aid in the filtering process. In case of sudden edges, we can expect initial filtering functions per se. Most of these edges can be due to disruptions caused and this will also initiate movement of the sediments within the boundaries. There will be gradual edges also. Most of these gradual edges can help in the movement of the materials between the ecosystems. There will be enhanced filtering and this can be witnessed along with the ecological gradients. Most of the gradual edges can be observed in natural settings and they will be diverse. The functions like source and sink can be considered as important aspects in the delineating stream waters. Most of the materials, energy and organisms can be supplied to the adjoining areas by these rivers. Some of the regions can act as sinks that can absorb materials and energy that can be generated from the adjoining landscape. It is to be assumed that the stream might act as both sink and source. Mostly, the location of the stream can be considered as an important feature in determining the nature and load it can carry. The genetic materials that survive on the landscape can be supplied by the corridors or through the corridors. This can be studied at the landscape scales. The sediments, energy, nutrients, groundwater and surface water can be effectively stored in the stream corridors. There are three important sink and source functions that can bring about vegetation in the floodplains. Some of the sources are water-borne organic matter and soil organic matter.

We need to consider dynamic equilibrium in studying the river ecosystems. If the anthropogenic intervention is not present, then also there will be changes in the water ecosystems. The stream corridors can exhibit dynamic equilibrium and hence there will be instantaneous stability. It is to be noted that the stability is based on several conditions that may vary. This can be called dynamic equilibrium. Most of the stream ecosystems can have natural correcting mechanisms. There will be external stresses and disruptions that can induce certain mechanisms in the ecosystem. This can be called a self-sustaining mechanism.

It is always hard to quantify the threshold levels that might exist in the ecosystems. It takes a lot of time for a river ecosystem to approach a steady-state condition as it is dependent on both internal and external factors. Once the anthropogenic disturbance is removed from the ecosystem or minimized, we can expect the rejuvenation of the riparian ecosystems and associated landscapes. The ecological stresses can be determined by the variations that might take place in the river ecosystems. Some of the ecological stresses can be classified as human-induced and natural. The land-use dynamics can be based on the hydraulic features and hydrological systems that might regulate the river ecosystems. These variations can alter ecology, habitat, and sediment transportation. The disturbance can be caused with equal intensity, duration, and frequency. These disturbances can occur in the associated ecosystems and stream corridors. The disturbances can vary within time and scale too. Natural stresses can also play a vital role in defining the river ecosystems. Some of the natural events like drought, temperature extremes, landslides, earthquakes, volcanic eruptions, floods along climate variations can be classified under natural stresses. All these events can bring about negative impacts on both function and structure of the river ecosystems. The disturbance that might occur in ecosystems can be based on resilience, resistance, and relative stability. The sedimentation and erosion can be classified as direct impacts that can be brought about by the impairment of ecology functions and structure. Most of the riverine vegetations can be assumed as resilient. In some cases, due to flood events, these riparian systems may be disturbed.

Dams can relatively damage the river ecosystems if proper management practices are not maintained. Mostly, the dam construction and its operation can alter the stream corridors. The ecology, sediment transportation and water quality in the impoundments can be damaged considerably. Most of these barriers can also affect the migratory organisms and resident biotic communities. The dams can slow down the migration of the aquatic species and their passage. These structures will also affect the stream' ecological functions and food chains of fishes. Even under naturally available conditions, the streams and river basins can exhibit high levels of stochastic variability.

The water diversions and channelization can be considered as important aspects of the river ecosystem variations. The channelization process can alter the stream ecology and hence the pool complexes can be disturbed. The diversion channels and Levees can also alter the riparian vegetations. If deforestation persists, the riparian vegetation can yield variable nutrients, temperature, and shading. If concrete is used in the riverine ecosystems, there will be reduced populations of amphibians, reptiles, and invertebrates. Water diversion can also affect the stream ecology and it will be based on the amount of water that is being diverted and the timing of diversion. The operation and design of the diversion structure should be managed properly.

## 10.7 HABITAT FRAGMENTATION

The habitat fragmentation (Figure 10.4) must be considered. In the case of fragmentation, there will be an extinction of species and ecology (Lenore et al., 2017). Mining is another aspect that can be considered to know the river ecosystems. The excavation of the bed gravel from the river ecosystems can result in damage to the local aquatic environments. It is worth noting that this will damage the benthic communities. In the mining process, there will be the usage of mercury and this will poison the river water and can damage human life also. Gold mining can be confined to certain regions but gravel mining can be witnessed in most of the areas due to construction activities. We can use coarse sand and gravel as building materials. Due to gravel mining, there will be habitat loss. The spawning grounds can be damaged due to gravel mining.

**Figure 10.4** Habitat Fragmentation.

**Source:** https://socratic.org/questions/58cf3abc11ef6b36bba33ce0

The mining at the surface can also result in stress on the riparian vegetation and landscapes. Due to surface mining, we can expect transportation of minerals, coal, processing, extraction, and exploration that might bring a lot of damage to the stream corridors. Most of the mining operations can disturb the nature of the watersheds. Pollution is yet another process that can damage all the species at a given landscape at any point in time. Toxic runoff can diminish the local vegetation and this will force the native species to shift to other tolerable locations. Most of the habitats will be severely affected by toxic precipitation and runoff. Most of the feeding organisms and bottom-dwelling species can be affected by iron precipitation because of the acid mine drainage. Mostly, a precipitate coating can occur at the bottom of the streams. This will damage the eggs of the biological species. Most of the fishes may not withstand the hostile conditions that can exist in the streams. Most of the contaminants generated from the industries and municipalities can exist for prolonged periods. The chemicals generated in agricultural practices can be classified as nonpoint pollution sources. We can also observe secondary effects due to sediments that are adhered to the agricultural chemicals. This will be due to herbicides and pesticides. Urbanization can produce several impacts on the local ecosystems. Most of the watersheds can be damaged due to increased population rates and unplanned urban management. Most of the riparian forests are being replaced by the urban bodies and this will lead to damage to the local water bodies.

Most of the river crossings can act as barriers to fish migration and other aquatic biotic migrations. The land use change and agriculture can be vital domains in explaining the anthropogenic inclusion and arbitration of the landscape dynamics. The exotic species can be introduced in case of limited resources in the landscape. There might occur inter-species and intra-species biological disturbances in the form of cannibalism and competition. The organization of community and size of the population can be considered as indicators to understand the natural interactions. The introduction and intervention of the exotic species may not completely damage the ecosystems per se.

We need to continuously assess the river ecosystems so that we can easily manage the ecosystems before actual damage. We need to calculate the river ecosystems in detail. There are some species that can act as indicator species. The response uniformity can be variable with the spatial dimensions. This deals with the response that can be projected by the indicator species against the environmental stresses. In some cases, the variability in the habitats and geographic locations can be important to the species that may elicit a response. The size of the species habitat can be important. The home range of the indicator species can be large. The species that can be minimal numbers cannot be used as indicator species as they may not elicit an appropriate response. The indicator species can bring about the appropriate response against the measured effect. Some of the fluctuations can reduce the ability or even restrict the ability to know the effects. There are some generalist species that can vary with the endemics. The woodpeckers and cardinals can serve as indicators to a limited extent. We need to consider the species sensitivity towards the environmental stresses that are under investigation.

We need to consider the cause-and-effect relationships. In some cases, we will be forced to stumble upon just the correlation studies. Most of the species that act as indicator species can be placed in the central domain of the food chains. This can be attributed to their high stability and responsiveness. The native species may exhibit narrow tolerance. The aquatic macroinvertebrates can be used as indicators of both riparian health and stream conditions. We can use benthic macro-invertebrates as best suited indicator species. These species can be abundantly available. Mostly, they will be placed at the central portion of the food chain of fishes. The sampling process can be easier. We can easily quantify the degraded conditions. Most of the environmental variables can be active for the short-term. The localized conditions can also serve as good indicators.

The field sampling may demand the cumulative effect of the qualitative and quantitative methods. Some of the environmental parameters like dissolved oxygen concentration, current velocity, water temperature, water depth and substrate composition can be considered. The aquatic hydrophytes that are present may also be considered for the analysis of the qualitative attributes.

Chapter 11

# Mitigation of Floods

## CONTENTS


11.1 Introduction .... 192

11.2 Climate Change Perspectives .... 195

11.3 Flood Characteristics .... 196

11.4 Hydroinformatics .... 196

11.5 Knowledge Management .... 198

11.6 Geopolitics .... 205

11.7 Environmental Impact Assessment .... 211

## 11.1 INTRODUCTION

The geostatistical tools (Figure 11.1) can be used in reservoir modeling. Geologists can use the fundamental concepts of Geostatistics in designing reservoirs. Geostatistics can bring about the necessity of the modeling process (Mallet et al., 1989). Any geological model can be linked with two important components. They are fluids and rocks. We can use geostatistical tools in the management of floods. The nature of the catchment in the local area can be important. There are some similarities that can exist between coastal flooding, urban flooding, and fluvial flooding. They share similarities.

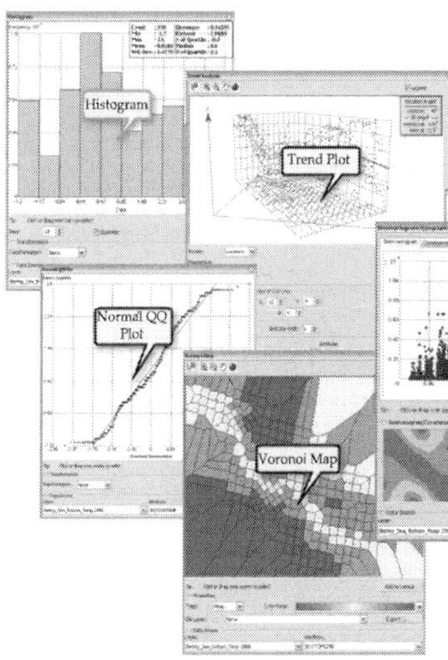


**Figure 11.1** Geostatistical tools.

**Source:** *https://desktop.arcgis.com/en/arcmap/latest/extensions/geostatistical-analyst/a-quick-tour-of-geostatistical-analyst.htm*

The flood disasters (Figure 11.2) can be disastrous and can induce several economic losses. The principles of flood risk management can be described by non-structural and structural measures. There are some catchment management plants that can be used to manage catchment and river basins. It is to be noted that the topography of a specific catchment is important and it will describe the drainage pattern of the rivers and streams. The natural causes of most floods are climate change, glacier melt, storm surge, landslides, and precipitation.

**Figure 11.2** Flood disasters.

**Source:***https://disasterphilanthropy.org/disaster/2019-u-s-spring-floods/*

The precipitation factor may involve snowmelt, hail, and rainfall. Landslides consist of seismic activity, erosion, and slope instability. The storm surge can result in variable tide levels (Ando., 1971). We can also expect enhanced groundwater levels and this will lead to saturated ground and runoff at the chalk catchments. The glacier melt can be due to volcanic action. The climatic factors can result in increased sea levels and may result in enhanced precipitation. Some of the anthropogenic causes are inadequate maintenance, inadequate integration, drainage, and planning controls along with the floodplain encroachment. The inadequate maintenance can be due to the sewer systems, culvert blockage and watercourses in the urban areas. The improper integration can be observed in the case of drainage systems and river systems. The drainage capacity will be limited due to siltation and urbanization processes. The siltation process can be classified as a natural cause (Graham., 1990). The drainage capacity can be hindered due to increased urbanization. The catchment area may not be planned perfectly both at the national and local levels and this will lead to floods.

**Figure 11.3** Encroachment of floodplains.

**Source:***https://sandrp.in/2019/08/30/aug-2019-yamuna-flood-floodplain-encroachments-exposed-again/*

The encroachment of floodplains (Figure 11.3) can result in lowered storage zones and this can be attributed to the construction of buildings on the floodplains. In some cases, the defense embankments that are meant to control the floods will fail. This will cause severe economic damage at the local level. Most of these dam failures can be attributed to improper piping, overtopping and it can lead to catastrophic disasters. It is always essential to demarcate the floods and their native setting. The floods can be long period floods and flash floods (Freddy., 2017). We can expect fluvial flooding if the river flow exceeds the discharge or we can call it as inbank flow. This will shift the water into the floodplains that are in the vicinity of the actual river flow path. This can be called as overbank flow. The flooding in the urban areas or urban flooding can be attributed to the less drainage capacity. This limited drainage capacity can be described by the irregular piped systems and urban watercourses. This can be linked with the interactions between the fluvial inputs and underground sewerage systems.

The coastal flooding can be linked with the overflow of the waters over the embankments. This can be common in case of high tide level situations. The flooding of rivers can be based on the native conditions. Climate change has increased the seal levels drastically so that even aquatic species are able to sense and migrate to more tolerable locations. The climatic change has increased the temperatures and humidity levels. Flood alleviation must be studied to know the actual flood risk situation. Most of the floods can damage the environment, social and economic tiers of the society. We can use a source and pathway model to know the risk that can be due to floods. We need to consider the technical uncertainties and aspects to know the damage and magnitude of the floods.

The floods can be reduced with structural measures. They are high flow diversions, river training, flood detention basins and dams. The high flow diversions can be achieved by the by-pass channels and add conveyance. The river training can be used in case of a lower floodplain, raised levees and enhanced conveyance. We can use flood detention basins to store water as in the case of off-line reservoirs. We can use dams as on-line reservoirs. The damage susceptibility can be reduced and this can be done by non-structural measures. We can consider mathematical and physical methods, floodproofing of buildings, and attempt development regulations, etc. The floodproofing of the buildings can help in the protection of the upper floors and lower floors. There will be an implementation of the development regulations that can help on a broad scale. We can use flood warning and forecasting systems to get the data on a real-time basis. The mathematical and physical models can be used to generate the flood risk maps and to optimize solutions that can moderate the damage yet to be caused. We can reduce the impact that can result due to flooding.

The early warning systems can be important to enlighten the local population about the disaster in advance. It is important to disseminate information about floods to the public. The problems that are associated with the floods must be discussed to avoid panic. Insurance agencies should be informed and this can help the local population to reduce their economic losses. We need to consider restoration of the natural resources also. It is important to implement flood insurance schemes that can help several poor and middle-class sectors. The tax structure should be modulated accordingly if someone uses the floodplains for construction activities. We need to plan the evacuation processes so that there will be redesigning of the buildings in the urban areas. Zoning is another important regulatory measure that can be used in the restoration of the forests, wetlands, and rivers.

The lowering of flood risk can be done with the aid of structural measures. In most cases, we can use all the available methods that can evaluate the flood risks. The modeling of river basins can be important in this case.

## 11.2 CLIMATE CHANGE PERSPECTIVES

Some of the problems associated with climate aberrations and their mitigation can be learnt by the improved portrayal of the natural variability.

There are several models that might not use the energy modulations at the interface of the atmosphere and oceans. Carbon cycle models and terrestrial models were emphasized and improved to match the native conditions. Feedback systems were used to know the actuality of the climate and vegetation. Rudimentary representation of the Ice and land was done. The treatment of water vapor was enhanced to bring about feedback systems. There will be a rise in uncertainty in cloud formation. We can also observe precipitation uncertainties. It is essential to understand the trends in thermal propagation over the landscape. It is important to know the climate scenarios at the regional levels. Most of the regional climate models were applied for water-related issues. The boundary conditions can be described by GCM models. There will be some rate of system errors that might develop during the calculation of the atmospheric dynamics. The GCMs derived results might not be completely perfect and hence this will reduce the quality of the regional climate models (Gabriel et al., 2011).

## 11.3 FLOOD CHARACTERISTICS

It is always important to opt for long-term observations over a region pertaining to precipitation. The trends in river flooding can be studied using statistical analysis. We can understand the effects of climate change on flood generation using the combined tools of hydrological models and climate models. It is important to use the meso-scale catchments. The impact of climate change on floods can be simulated using quantitative analysis. We can use global atmosphere ocean circulation models in climate change research to know the reservoir characteristics. There are several methods that can help us know the climate forcing and assess it with our mathematical models. It is important to opt for simple interpolation approaches, stochastic weather generators and empirical downscaling processes. We can consider the unsaturated zones and soil surface that can determine the runoff processes (Takemura et al., 2013).

## 11.4 HYDROINFORMATICS

Modeling can be considered as a central idea of hydroinformatics (Figure 11.4) (Roger et al., 2000). The modeling process can consist of certain steps that are well defined. First, we need to consider the physical laws. These laws can be applied using mathematical algorithms. It should be able to get replicated so that anyone can represent it digitally. We need to solve the difference equations with the native conditions and boundary constraints.

We need to design and test the numerical methods that can simulate the native conditions and processes. Some software like CAD and GIS can be used in order to simulate the processes that might occur over the surface of the earth.

Most of this software can utilize data linked with watersheds to make models. The model formation can be based on a series of steps. Firstly, we need to decide the actual use of the model. It is important to review the existing data. If data is insufficient, we need to procure data that can represent all the natural conditions. We need to consider consistency tests. The parameters that can calibrate the model should be determined. Then we need to confirm the model. The modeling software has brought several improvements in modeling coastal waters, estuaries, and rivers. Data-driven modeling can be based on the physics of natural systems.

**Figure 11.4** Hydroinformatics.

**Source**: *https://www.researchgate.net/publication/335951616_Urban_Hydro-informatics_Past_Present_and_Future/figures?lo=1*

This modeling process is quite different from the physically-based models. We can use artificial neural networks that can be used in place of data-driven models. We can use biological neural systems to process this information so that we can easily simulate the natural systems as and when needed. We can use the nearest neighbor model with the assumption that the closest points can be classified with similar classification techniques to the distant ones. The fuzzy rule-based systems can be used in the modeling process. It consists of both membership functions like input and output. Most of the fuzzy rules can be combined with the inference engine. We can also use genetic programming techniques in this case. Genetic programming can be used to know the evolutionary rules that can be generated with the input and output sets. We can use model trees and decision trees. These decision trees can be demarcated into roots and trees. We can use support vector machines also. In this case, we can use an approximating function to verify the training sets. This can be done with the statistical learning theory. We can use the fuzzy rule-based system in the prediction of the precipitation events. The uncertainty in groundwater availability can be based on fuzzy logic. We can use a fuzzy rule-dependent system to manage water in the water-scarce areas.

## 11.5 KNOWLEDGE MANAGEMENT

Knowledge management can be considered as a crucial factor in the water sector (Regis., 2017). We need to evaluate the knowledge management mechanisms as and when needed. Knowledge mapping is needed to evaluate the actual water situation in the landscape. We can use multimedia communications also. There should be a collaborative process and knowledge sharing. We need to manage content and documents. The encapsulated knowledge systems can be based on the open modeling systems and data-driven modeling systems. We need to understand the learning process at both organizational and individual levels. We need to use hydroinformatics that can serve as a link between natural science and physical science. The information and communication technology can be integrated in relation to the social context. The hydraulic models and hydrological models can be associated with the time series and can be represented easily. This can be applied at several cross-sections also. We can define these cross-sections as representative cross-sections. These cross-sections can be linked with the water level limits across the watercourses. We can easily classify the watercourses according to their levels.

The flood events can be considered as normal reactions presented by nature. The floods have existed even before anthropogenic activities initiated them. it is to be noted that human interference in the flood generation process must be mitigated, reversed or compensated to bring about certain changes or balance in the water storage systems. The flood strategy must be designed in such a way that the whole river Basin is covered. With these strategies, we should be able to manage several actions pertaining to river rejuvenation and its maintenance. We need to link land and water resources accordingly. We need to consider certain trends followed by natural hazards at a particular region using archival data available from the standard libraries. It is always essential to opt for management options rather than defense reactions in the case of flood mitigation. The mitigation of floods may not be confined to the prevention of floods but several other factors too. We should also consider rare events as part of the flood mitigation process. We need to attract transnational efforts to restore the rivers and the flood zones by protecting floodplains. We should frame strict laws about the floodplains being encroached by the builders. Humans may utilize floodplains for various activities. There should be certain measures to regulate flood-associated problems. Some of the problems that can be tackled with appropriate flood mitigation strategies are pollution, ice flows, landslides, mass deposition, erosion, disruption of the sewage network, increased groundwater tables and flooding. There will be a need for non-structural measures and mitigation measures to bring about sustainable solutions to the problems that are relevant to water bodies. It is always important to reduce the flood risk and minimize human vulnerability towards floods. Most of the structural measures can be categorized under defense structures. These structural measures can be classified as essential elements to protect human health along with goods and services. It is always worth noting that no flood production management practice is absolute. In some cases, the flood mitigation strategies may appear as exaggerated risk management.

The flood warning and forecasting systems should be installed in order to maintain standards in the flood mitigation protocols. The correct response and the timely intervention along with preparedness will bring about the effectiveness of any flood management strategy. It is a part of the responsibilities to provide this information in advance to the public and industrial sectors well before the flood event. The safety preparedness and measures to be taken should be inserted in all the hazard management protocols of the public and private institutions or companies. One should take into consideration the solidarity in addressing the water management problems at a specific area or any region in question. Generally, these strategies come with three main steps. The first step can be called the retaining step followed by storing and later draining step. It is important to collect rainfall through water harvesting methods in both urban and rural areas to reduce surface runoff and anchorage infiltration. This infiltration can improve the groundwater levels in the local areas.

In most of the areas, flood prevention was classified as a precautionary principle per se. We need to deal with certain compensation systems that can help the victims in flood situations. It will also improve the economic status of the victims who lost everything in the flood events. We need to focus on insurance solutions at both the public and private levels, to subside the dangerous and disastrous effects of the flood basin damage and hence the solidarity will be advanced. It is important to trace flood-prone areas well before a flood event. The adverse effects caused by this event can damage the whole Terrestrial ecosystem and aquatic systems in some areas and it is dependent on the scale of flood events. the soil pollution and water pollution will be much higher than normal conditions. The contribution of environmental conditions to the flood events must be investigated in order to know the actual anthropogenic contribution towards this disaster event. There are four primary mitigation approaches that can be adopted by the local authorities. They protect, accommodate, retreat and avoid. In the case of a protected approach, we need to avoid floods that reach developed areas. This can be done with the help of dykes installed in the river channels. We can also construct dams to store water and it will help us in reducing the risk whilst peak flows of rivers. In the case of an accommodate approach, we need to adapt certain land use plans to reduce damage. With this approach, we can raise buildings higher than the flood levels prescribed by the local authorities and municipalities. We need to construct buildings with specialized materials which can render the building less vulnerable to flood situations. In the retreat approach, we need to think about withdrawal of the floodwaters mainly in the hazard areas. In case of flooding of the local landscape, there is a possibility after converting these damaged lands into natural spaces and parks. In the case of a reward strategy, we can avoid certain hazard zones from the building activities. There should always be Limited growth in the hazard areas and vulnerable zones. Sufficient space should be left for the rivers to flow even in the urban areas. In urban areas, floodproofing is related to non-structural and structural changes that are being adopted in the building construction to reduce flood damage. The floodproofing can be done through elevation, dry floodproofing and wet floodproofing. Increasing the elevation of buildings well above the flood level can be a reliable method in the floodproofing process. Buildings can be made watertight in order to avoid water being entered into the building at this can be called as dry floodproofing. In some cases, some of the portions of the building will permit floodwaters and then allow their exit without causing any damage to the building structure.

One of the conventional flood mitigation plans was structural river training plans. This method was a general flood management approach that was adopted by the management authorities of flood plains. In recent times, there were several non-structural approaches that were proved to be effective in flood risk management. Then non-structural approaches were better than the structural approaches. The risk of floods can be drastically reduced by minimizing the vulnerability of sensitive areas. The damage analysis can bring about several alternatives that are needed to protect the public at large by optimizing the size of the structural measures. The cost of risks should be quantified in this analysis. The risk that might happen to the structures can be translated into tangible and intangible costs. Most of the tangible costs then explain the damage that is being caused to structures and properties, repair costs and loss of businesses. The intangible cost can include damage to the environment, social unrest, loss of human life and psychological trauma caused due to flood events. Intangible and tangible costs must be justified based on the vulnerability status of the specific area. We cannot quantify all the benefits of the project being started without taking into consideration risk cost. This concept has become a normal mandate for several flood mitigation systems. Some of them are storm surge protection, spillways and dams along with storm sewer systems. It is always important to consider the construction of hydraulic structures based on risk forecasts and mitigation. We need to understand reliability analysis while starting a project for designing a project. In the case of flood mitigation plans, there will be several practical measures that are to be considered to minimize the flood damage. Flood mitigation plants should include certain important issues that are to be addressed. First, we need to know about the best implementation method to regulate flood situations. It is important to select appropriate locations so that reflection medication facilities can be installed accordingly. The size of the facilities should be justified based on the available local resources. We need to consider appropriate methods for the successful operation and maintenance of these flood mitigation facilities under investigation. The river training plans along with flood damage analysis can be considered as important aspects of project planning for both existing and planned projects.

The key component risk-based analysis that can be used in the flood management protocols are 1. reviewing of archival data about the frequency and magnitude of the floods and their respective analysis 2. the simulation of flood control at the upstream portions of the reservoirs 3. procuring historical data of the damage caused by previous flood events 4. field survey of the flood damage. We need to collect the data pertaining to the hydrodynamics of rivers. the flood mitigation options can be analyzed accordingly. Simulation models should be able to protect the flood levels accurately and define the standards appropriately. The cost benefit and economic analysis should always be taken into consideration to dissipate risk-based analysis based on flood damage.

The flood mitigation studies should investigate the critical zones of flooding and should always mention the methods needed for damage control caused by floods. The analysis of flood frequency is always needed to know the magnitude of the floods. It is always important to consider the intermediate catchments and upper catchments to know the risk-based flood analysis. We need to simulate the Reservoir flood control strategies to know the actual situation of the upper storage portions of the reservoirs. The flood hydrograph can be used in this case to know the actual situation that may exist in the downstream area and case of return periods. The initial conditions of reservoirs must be recorded and it should be studied to speculate the condition of the attachments with the help of a flood hydrograph. Then inundated areas can be the best hotspots to estimate flood damage and it can be considered as a key component in flood related studies.

It is human that without people there is no hazard. It is always important to increase the Awareness of flood preparedness to combat the damage caused due to flood situations. Maps can be used to identify and observe the flood risk areas at all levels that will be part of the flood management Action Plan. These maps can be prepared using historical information along with the hydraulic elevation of the landscape. The water management authority should take into consideration the flood mapping protocols accordingly. The flood maps can be prepared with essential information. we need to measure flood flow and flood levels and information from the previous flood events should be procured before preparation of the maps. It is always essential to collect topographic information of a specific area that is to be mapped. We need to understand detailed flood maps so that the population that is living in a particular area can be taken care of when flood events arise. The vulnerable individuals and vulnerable groups must be taken into consideration in order to frame equation routes so that the flood maps can be prepared in a timely manner. Flood maps can be integrated with the height of the floods well in advance. Flood hazard maps are very much essential for the public at large. In the production of flood hazard maps, we need to know about flow velocity, depth and flow extent. Even communities residing in a particular area can produce maps based on individual requirements and community interest. We can use the seasonal calendar to know the information about mountains and their effect in vulnerable areas. A seasonal calendar mostly provides information about the meteorological conditions and hydrological cycles. We can get fundamental information about rainfall with the archival data available from the weather stations. Flood awareness is always important to educate the communities so that they are prepared for the flood event in advance. It is quite essential for the communities that are at risk and that means communities are living in flood plains or associated areas. Generally, these sorts of works should be taken care of by non-government organizations and local authorities that are dealing with the public at all times.

Public meetings can be conducted regularly at certain times of a year to review the options about the flood management and reduction of the damage that might occur during such an event. We can use booklets and posters produced by the local community to keep reminders of all the events that had occurred earlier and take proper preventive measures to avoid economic losses and human life. We can also conduct educational activities in the colleges and schools so that the information can be dispensed at large in most of the communities. Drama and theatres can also be performed to explain the situations that may and entangle the society in case of flood events. Impending conditions that may lead to floods can be forecasted by the authorities. Flood forecasting mainly requires the local hydrological and meteorological conditions that control the local rainfall. We need to consider the hydrometric network before relying on the forecasting system. The hydrometric network comprises water level and rain gauges at ground level so that we can get accurate and real-time information when needed. We need to rely on international data and forecasts including weather radar information and weather satellite information to bring about flood forecasts effectively. The flood forecasting operation can be associated with the nearby countries and this can be prominent in the river basin that is being shared. It is worth noting that government services should always spend some part of the budget on flood forecasting systems to protect the public at all times. We need to establish automatic hydrometeorological stations that can be linked with the Central command centre by telemetry and radio. We need to design field forecasting software that can be linked with observation networks at ground stations. The primary components of any national flood forecasting system can be based on several factors. We need to collect data in real-time and it is always essential to predict the severity of the floods. We need to consider specific flooding levels. We need to prepare warning messages that can disseminate information in real-time. The dissemination and communication of messages are very important in the early warning systems. We need to interpret and predict the possible impacts of floods on local communities. It is always essential to respond to the queries given by the local communities instantaneously to reduce the distress caused by the flood early warning information systems. We need to review the warning systems and their capabilities periodically so that the accuracy of the systems can be met. Different warning systems are different from forecast and the flood warnings can be given to certain parts of users only. We need to correlate emergency personal and operation teams so that the state of readiness will always be maintained without any problem. We need to warn the public about the spatial and temporal extent of the flood event. Most of the warnings can bring distress to the public as they will impact roads with heavy traffic and the people who are relocating.

In exceptional cases, we need to prepare emergency procedures and evacuation procedures well before dispensing information about the flood forecast. It is important to always deal with preparation time so that we can easily inform organizations and individuals with a state of mind that can be prepared for flooding events. Early warning systems that are installed in local communities can always save a lot of lives and economies. The flood warnings must be quick and clear so that most of the public can understand and propagate the information. We can use pictorial codes so that the public at large can understand easily. We need to review if the warning messages are standard or not. If flood warnings must be obvious. It should also describe the areas that are at high risk. The maximum water level that is possible during a flood event must be stated and the time of flood should be mentioned. The duration of the flood can be taken into consideration and stated wherever possible with help of local communities. There are certain sections of society that don't care about the flood warning systems and messages and are ready to take risks. Some of the other associated priorities may deviate the public from listening to warning messages provided by the authorities. The local weather conditions and actions by the officials in the local area may differ from warnings stated by the flood warning systems as this will lead to confusion among the public seeking actual information. Some people might even express aversion to the messages and often ignore advice given by the local authorities. In some cases, people might not respond to the warnings and hence there is no value in improperly getting information about the anticipated floods in that area. The physical and mental capacity of the public may not suit the valuable information provided by the flood management authorities. It becomes a duty of the local authorities to sustain information and make sure the public at large respond to it in a proper way. some people might not be ready to accept the fact unless there is a financial loss or loss of human life are both. The flood warnings must be generated from a single reliable source. This will avoid confusing the public who are seeking for accurate and reliable information. The warnings that are to be delivered may not be controlled by the management authority at some levels due to cross-communication among the public. In this case, the potential users may not get timely information and may be at the behest of rumors. It is always essential to review the communication systems regularly. At the community level, most of the warnings will be taken by the individuals and observed by them resulting in proper action from their side. We can opt for media warnings and general warning indicators at community levels. We can make use of community leaders to propagate information in a better way at emergency modes. We need to maintain automated telephone warning systems you communicate to the public as and when required. The trained individuals are necessary for the forecast management authority to receive reliable information.

## 11.6 GEOPOLITICS

Geographic features can play an important role in determining the geopolitics of a particular area. We can assume that the geographical conditions as building blocks of any analysis pertaining to local politics and in this case geopolitics (Petr., 2006). The geography of a country can decide the water resources and they impregnated geopolitics (Figure 11.5) thereof.

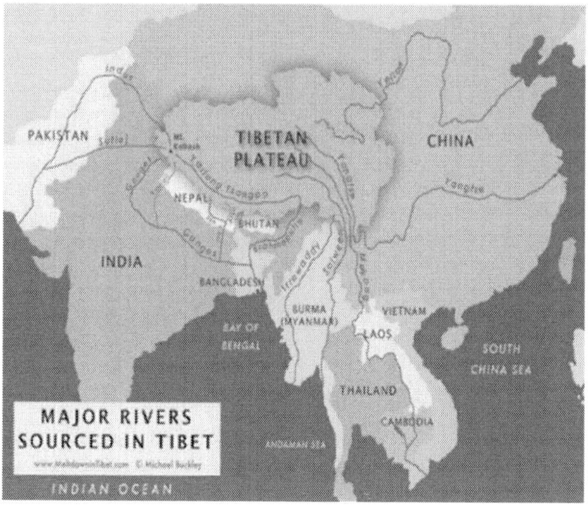


**Figure 11.5** Geopolitics.

**Source:***https://www.chinawaterrisk.org/resources/analysis-reviews/geopolitical-risks-transboundary-rivers/*

The impact of geopolitics in any country is very important to determine the local water resources and their progress all along the landscape. Most geopolitics pertaining to river environment and the marine environment get surfaced during war conditions (Gertjan., 2004). one of the best examples to describe this is the Pacific War of South America in 1879 that lasted until 1883. This war was initiated to gain power in the Southern Pacific region. In this war, Bolivia lost its ground and still, there are a lot of fights pertaining to remnants of the war. The invasion of Crimea by Russia is a similar sort of invasion that is to be specified when dealing with your politics. It is essential to understand that waterways can provide us economic feasibility as far as the import and export of goods is concerned (Jacques., 2000). The Mississippi River system in the United States can be considered as one of the strategic riverways that controls the economy of the developed countries.

**Figure 11.6** Ohio river.

**Source:***https://www.circleofblue.org/2013/world/choke-point-u-s-water-energy-and-the-ohio-river-valleys-new-course-2/*

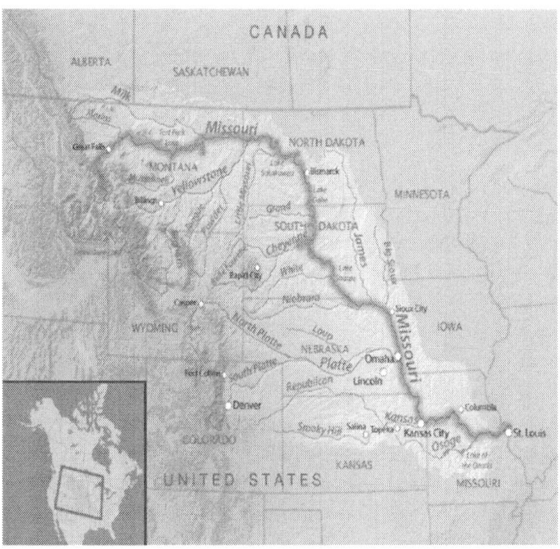


**Figure 11.7** Missouri river.

**Source:***https://www.chinawaterrisk.org/resources/analysis-reviews/geopolitical-risks-transboundary-rivers/*

Ohio and Missouri rivers (Figure 11.6 & 11.7) can flow into the Mississippi making it a large river system. In this river system, we can easily navigate and it is considered a friendly river environment that can easily get emptied into the Gulf of Mexico (Alan., 2019). Any imports and Exports that are generated from Appalachians and Rockies can easily be shipped to Europe using this River system. In order to acquire this river system, the United States has waged war against Mexico. This has resulted in an increase in the buffer zone related to the Mississippi River. Most of the rivers can be attached with geopolitical power (Jarvis., 2011). We can consider the Nile river as a better example to explain geopolitics (Zainidin., 2011). It is worth noting that at least 85% of Nile river water originates from Ethiopia, Sobatrivers, Atbara and Blue Nile rivers. The Blue Nile river can be considered as a major contributor to the Nile river and it accounts for 60% of water. Anyone who initiates steps to alter the course of the upper segment of the Blue Nile river can affect several territories downstream (Marwa., 2011). It can even impair the accessibility to the waters of the river Nile. The Grand Ethiopian Renaissance Dam is creating such a vortex of geopolitical situations in both Ethiopia and Egypt. Egypt has waged war against Ethiopia in order to control the Nile (Eva., 2003).

In the case of Syria, the scarcity of water resulted in large conflict. There was a drought in Syria from 2000 to 2010 and it had reduced livestock and Agriculture making Syria a country with less food supplies (Kevin., 2001). This is the reason why the Islamic state has regulated this territory for a certain period after the drought conditions. The Islamic state has taken over this territory because of socio-economic devastation due to drought conditions and this will lead to worldwide attention on Syria (Duncan., 2020). The long-lasting famine because of drought conditions in Syria has made it vulnerable to attracting the source from radical groups. Most of the geopolitics imperatives can be dependent on the accessibility to water resources. The strategic wellbeing of a country will be based on waterways and access to the local water bodies without any problem (Graham., 2011).

China has at least 10 major rivers that are being shared with surrounding countries with the least formal agreements. The stress and scarcity of water have drastically increased the political tension in China's territory for decades (Leszek et al., 2019). The dam constructions in the China territory have escalated a lot of problems with the neighboring countries. Qinghai-Tibetan plateau covers around 2.5 million square kilometers and it is considered as the roof of the world and some would call it as the third pole. This plateau acts as a shelter for two most of the freshwater reserves apart from the North and South poles. Most of the rivers of Asia that travel from west to east for being generated from this plateau Such as the Yangtze, Yellow, Mekong, Salween, Irrawaddy, Brahmaputra, Ganges and Indus. Except for Yangtze and Yellow, the rest of these rivers can be classified as transboundary rivers. The Mekong river extends to several countries and it is charged to South China seas, Andaman and the Bay of Bengal. This can be considered as one of the first water resources for Vietnam, Cambodia, Thailand, Laos, Burma, Bangladesh, Nepal, India and Pakistan. Most of these transboundary rivers are the central points of political tension globally. It is to be known that at least one-third of China constitutes the International river Basin. This International river Basin is being shared by at least 18 rivers. Indus river floors at least 3200kilometers. It flows from the Tibetan plateau. It flows via Jammu and Kashmir through Northern parts of India before entering Pakistan followed by the Arabian Sea. The Indus River Basin can be about 1000000 square kilometres. This constitutes about 10.2% of the watershed in China territory. Brahmaputra river extends to 3800 kilometers. It basically generates from the Northern part of Himalayas and reaches Assam in India followed by Bangladesh. It joins the Ganges and reaches the Bay of Bengal and hence forms the Ganges delta. This river basin is approximately 650000 square kilometers. The Ganges may extend to 2500 kilometers. It starts from the western part of the Tibetan plateau and reaches the east and south parts of India. It also enters Bangladesh and joins the Brahmaputra before dispensing into the Bay of Bengal. The Ganges River Basin is approximately 12500 square kilometers. Brahmaputra, Ganges and Meghna rivers constitute 19% of the watershed over Chinese territory. The Mekong river extends to 4900 kilometers. It starts from the Eastern part of the Tibetan plateau and reaches the Mekong delta. During its trip to the Mekong Delta, it touches Vietnam, Cambodia, Thailand, Myanmar China. It constitutes at least 21% of water shade in China. Irrawaddy river can extend to 2100 kilometres. IATA generated from the Tibetan plateau via Burma reaches the Andaman sea. It spreads upto 411000 square kilometres: this constitutes 4.6 % of the China watershed. Salween river extends to 2400 kilometers. It extends from the Tibetan plateau to the Eastern part of Burma and then enters the Andaman sea. It constitutes 52.4 3% of the Chinese watershed.

Water scarcity has become a common word to describe any country›s situation these days. This can be attributed to ongoing water insecurity. In the Asian continent, Pakistan is designated as a country with a severe water security risk. Cambodia, Thailand, Bangladesh, India and China were classified as countries with a high water security risk. These statistics were given by Maple croft. This index mainly takes into consideration the Investments, operations and supply chains that are being maintained by multinational companies. There are four important areas that are being considered by this index. The dependency of the economy on the water will be considered with top priority while making any speculations. The relationship between supply demands and water will be taken into consideration. As part of this index, the amount of renewable water and the external supplies will be estimated. The fundamental access to sanitation and drinking water has become part of this index. In 40 years, the global population is expected to rise by 50% and hence the water demands will also be increased in a geometric proportion. This will also create a window for a lot of diseases due to lack of purified water and even its unavailability.

There are several theories explaining transboundary water-related geopolitics but none of them was satisfactory as far as the river basins are concerned. The political relations between domestic politics and riparian societies are becoming weak as the days get progress. There are a lot of economical insufficiencies and Limited regional interventions that are making water-related geopolitics a nightmare. This can be attributed to non-critical approaches towards power, scale and geopolitics. Oil was considered as an important aspect that can initiate in present-day situations but it is overwhelming to say that oil was replaced by water. Water Wars are in place in most of the territories as we know.

The water disputes were linked with regime survival, political legitimacy and domestic stability. This can be termed as Hydro geopolitics. There are several discussions on geopolitics and energy over the last decades in order to enhance the energy security of the world. Climate security also joined with energy security issues while debating about environmental situations. Still, there are many geopolitical implications that are being controlled by key factors. The melting of ice, territorial conflicts and water conflicts are some of the factors that play an important role in declaring geopolitics in particular countries. Climate change has resulted in water insecurities over many territories. Greenland and the Arctic are some of the zones that are to be considered while discussing geopolitics at large. The increasing global temperatures resulted in the melting of most of the polar ice caps and this brought about an increase in the sea level. At least 10% of the seas what are the degrees according to National Aeronautics and space administration reports. These readings were supported by laser altimeters. Most countries are very interested in extracting minerals, natural gas and oils. The Arctic zone supports 30% of natural gas and 13% of oils that have yet to be excavated. These attempts by countries can bring about detrimental effects on the freshwater resources our planet provides.

The passages were opened in Northwest and northeast frontiers and this can be attributed to excess melting of ice. This can be a central concern about most of the geopolitics all over the world. There were tensions related to water issues among the United States, Russia, Norway, Denmark and Canada. The Grand land was considered a favorite centre for military operations between North America and Russia. Greenland was seasonal by the United States to control North America.

## 11.7 ENVIRONMENTAL IMPACT ASSESSMENT

The environmental impact assessment of any Reservoir construction can be classified as an economic strategic perspective that can be linked with the economic development of any country. We need to consider watersheds in order to know the multidisciplinary nature of the Reservoir is under investigation through proper Collection of data from socio-economic studies, environmental studies, biological experimentation, local geology, etc. Reservoirs can be linked with watersheds and this will be called as biogeo physiographic boundary in geological and hydrological terms. If an entire area is being drained by the River system then we call it a watershed. We need to take into consideration the functioning and dynamics of Reservoir ecology so that we can know the proportion of anthropogenic activities that might cause a disturbance. Most of the water systems can be evaluated using morphological, genetic and physical parameters. Watershed conditions mostly will be based on anthropogenic activities. We can classify the severity of impacts based on several quantification studies along with quality parameters. We need to know the impact dimension, spatio-temporal component of the reservoirs. The magnitude of the importance must be demarcated with ecosystem relevance. The actual nature of the impact must be estimated. We can use the Restoration economy in order to bring about certain advantages from Reservoir construction. Most of the reservoirs can be classified as recipients of natural and anthropogenic products that might have arrived from the watersheds. This has been a fundamental principle to define water shirts based on environmental impacts.

We can evaluate reservoir construction based on technological dynamics, ecological dynamics, connectivity and degree of organization. We need to consider the connectivity aspect of the watershed and the reservoirs. This is mandatory because the artificial Constructions are being made on the naturally occurring landscape and may influence Ecological complexities. Watershed can be considered as a fundamental unit environmental impact assessment of any reservoirs. It is always essential to work complete and systematic analysis of watershed ecosystems even before the site selection process is being made. The water budget and hydrological component can regulate watershed cycles and their correlated dynamics. It is essential to observe and measure evapotranspiration, runoff and precipitation of the site selected. Water storage becomes an essential part that is to be evaluated before the construction of the reservoir. We need to consider hydrological pathways along with river ecodynamics because most of the biotic parameters like biomass, productivity and distribution will be affected because of this artificial construction.

# INDEX


**A**

Aeration 119
agricultural activities 36, 44
agricultural rivers 92
Alluvial bend 98
alluvial channels 156
alum 120
ammonia 134, 144
animals 92, 106
aquatic ecosystem 181, 182
aquatic plants 180, 181, 183
atmosphere 196

**B**

biological systems 135, 137
bleaching powder 120
braided rivers 98, 99

**C**

carbonate 51, 54, 56, 65
cement slurry 7
Chemical remediation 120
chemicals 118, 120, 123, 128, 129
civilization 36
Clay particles 7
climate 36, 43, 45, 47
climate change patterns 15
coalbed 51
coal mines 69
coastal flooding 192, 194
computer-based simulations 50
corridor 178, 180, 185
crude oils 4
cyanobacteria 138, 139

**D**

Darcy's law 5, 6
Deltas 153
detergents 69
Dissolved oxygen 124, 125
Disturbance theory 100
domestic needs 36
drilling 50, 57
droughts 16
dyes 116

**E**

earth's surface 92
ecosystems 14, 16, 32
ecosystem sustainability 14
Eichhornia 139
environmental legislation 153
erode 92
erodibility 98
erosion 92, 97, 98, 101, 105, 106, 107, 108, 109, 111, 112, 113
Eutrophication 141, 145, 146
evaporation 68, 76, 77

**F**

fauna 181, 182, 185
ferrous sulfate 120
floating plants 139
flood control 132, 143, 144
flood disasters 192
floodplain 38
flood risk management 192, 201
floods 14, 16, 32, 192, 193, 194, 195, 196, 198, 200, 201, 202, 203, 205
Flow regimes 4, 8, 9
fluvial flooding 192, 194
fluvial geomorphology 17
fluvial morphology 156, 168, 174
fluvial systems 157, 158, 159, 160, 163, 169, 170, 173, 174, 175

**G**

gas 50, 51, 52
gasoline 121, 123
geochemistry 50
geology 50
geomorphology 92
geophysics 50
geostatistical tools 192
glacier 192, 193
grease 69
groundwater 16, 17, 18, 27

**H**

hemispherical flow 5
human encroachments 14
hydraulic gradients 18
Hydrological cycles 68

**I**

inorganic chemicals 69
inorganic pollutants 116
interference 8
irrigation 132, 143, 144

**K**

Knowledge management 198

**L**

lagoons 153
lakes 153, 157, 164
landforms 150, 158, 175, 176
land resources 14
Landslides 193
land surface 68
Linear flow 5
lithology 36

**M**

matrix 178, 179
minerals 92
mitigation 195, 199, 200, 201, 202
mobility 8
molecular oxygen 124
mosaic 178
mud filtrates 7

## N

Natural drainage systems 119
natural ecosystems 36, 41
natural resources 36, 39, 40, 41
natural stream 119
Natural waters 132
neutron porosity 56
Nutrient loading 136
nutrients 36, 38

## O

oceans 150, 153, 157, 158, 159
oil 69
organic chemicals 69
Organic contaminants 124
organic pollutants 116, 124
Oxygen 183

## P

packer failure 8
partial penetration 8, 9
patch 178, 179
pesticides 69
petroleum engineering 50
petrophysics 50
phase separation 8, 9
photocatalytic reactions 124
photochemical reactions 124
Physical remediation 119
phytoplankton 136, 137, 141, 144
pollution 36
poly aluminum 120
population explosion 14, 15
population growth 14
porous media 4
production 50, 52, 53, 57, 59, 65
Pseudo steady-state flow 4

## R

radial flow 5, 6, 8, 11
Rainfall 68
remote sensing technology 92
Reservoir 3, 4, 5, 8, 10, 11, 12
reservoir analysis 56, 59
Reservoir engineering 50
reservoir modeling 192
Reservoirs 132
Rills 19
river corridors 180
river ecosystems 36
River management 14
river network 155
River pollution 116, 117, 123
river resources 14
river systems 92, 95, 96, 97, 101, 105
rock material 161
rocks 18, 21, 27, 92

## S

salinity control systems 70
Salinization 142
Salvinia 139
sandstone 50, 54, 57, 59, 64
seas 153
sedimentation process 92, 102, 105, 107
sediment control systems 70
sediments 36, 40, 92, 94, 96, 97, 99, 101, 102, 103, 105, 106, 107, 108, 110, 111, 113, 114, 156, 157, 158, 159, 161, 162, 165, 166, 167, 168, 170, 171
seiches 134
semi steady-state flow 4
skin zone 7, 8

slope 36
snow 68, 73, 78
soils 92, 106, 113
statistical analysis 196
storm surge 192, 193, 201
stratified layers 8
stream biota 181
streams 178, 182, 183, 185, 187, 189
stream water 178
sunlight 68, 124, 129
superposition 8
sustainable production systems 40

**T**

transient flow 4, 7, 8
tropical reservoirs 134

**U**

urban flooding 192, 194

**V**

vegetables 92

**W**

wastewater 70, 71
water aeration 120
water distribution 68, 71, 72, 89
water flow 38, 46
water pollution 121, 122, 123
Water quality 141
water resources engineering 68
water retention 16
watersheds 178, 179, 189
watershed scale 179, 180
water supply 68, 70, 71, 77, 85, 88
water systems 4
wellbore damage 7